THE BOOK OF THE UNIMAT

Fig. 1.7 Ideal summer workshop! The author at work on the patio in the shade of an umbrella. Lathe is attached to workbench fitted to popular folding Workmate portable bench. Lead comes from house behind.

THE BOOK OF THE UNIMAT

including SL and Mk3

D. J. Laidlaw-Dickson

A guide to its operation, accessories and possibilities aimed primarily at the user as a 'first timer.'

**Model and Allied Publications
Argus Books Limited**

14 St James Road, Watford, Herts, England

Model and Allied Publication
Argus Books Limited
14 St James Road, Watford,
Herts, England

All rights reserved. No part
of this publication may be
reproduced in any form or by
any means without the permission of Argus Books Ltd

First published 1977
© L-D. Editorial & Technical Services Ltd

ISBN 0 85242 591 0

Typeset and printed
in Great Britain by
G. A. Pindar & Son Ltd., PO Box 8, Newlands Park, Scarborough

ACKNOWLEDGMENTS . . .

Whilst this book makes no claim to be the official publication of the manufacturers I would like to express my very great appreciation of the help provided by the parent manufacturing company Messrs. Maier & Co of Hallein in Austria and in particular Herr Unterberger who handles their technical translation work. He has been specially helpful in giving me advance information of the Mark 3 lathe, with early sight of prototype accessories, a splendid range of additional photographs, and a most cordial welcome to the Austrian factory, where I was shown Unimats and the other larger lathes made by the company coming off the production lines in their hundreds. It was a revelation to see the constant quality control, with every worker equipped with extensive measuring equipment, right up to the ultimate view branch examination when duly stamped each lathe was trundled out and labelled for despatch.

The British distributors Messrs. Elliotts have also continued the good work, letting me have an early delivery of the new model, plus accessories at the very head of the queue.

I would thank also the help given by my colleagues, and in particular Martin Evans, the former Editor of *Model Engineer* who has provided solutions to all kinds of problems, in particular how to do the *impossible* things on the little Unimat.

Acknowledgement is also given for permission to reproduce drawings from the official publications, which are so clear that they could hardly be improved upon. Most of the photographs are of my own set-ups which I have taken; others—the better ones!—are mainly from Herr Schwaiger's camera in Austria. Herr Schwaiger is another Unimat enthusiast from the Hallein works who happily combines his official duties of staff photographer with highly skilled model engineering. Just to show it is possible he has built up a Stuart No. 10 from standard castings and finished it completely on the little Unimat.

CONTENTS

Foreword

I THE UNIMAT LATHE
Why a Lathe? How it works. The Parts of the Lathe. Suitable Mounting. Parts included in standard package. First minor improvements. Operating procedure. Upkeep and maintenance.

II ADDITIONAL EQUIPMENT AVAILABLE
Three jaw chuck or four-jaw? Drill chuck. Machine Vice. Live centre. Using the three jaw chuck, reversing jaws. Four jaw uses. Chalk mark setting-up. Care of chucks. Automatic feed attachment or self-act. 'League table.'

III NEW UNIMAT MARK 3
Need for complete re-vamp. Follows 'big brother' pattern. Main changes. Benefits obvious and less obvious. Increased size and power wider speed range. Incompatible nature of many accessories with original SL type.

IV MARK 3 ACCESSORIES
Improved three-jaw s.c. chuck. Improved faceplate with T-slots. (Standard faceplate in alloy as SL) Completely new compound slide with 0–30° dial. Fine feed to vertical drill. Vertical column now an extra. Flycutter. Mounting bridge for vertical column. Cross-slide mounting alternative for column. Improved threading hob now gives longer threading run. Indexing attachment with 24, 30, 36, 40 division plates. Three point steady more elegant. Self-act has built-in fixing and installs as 'designed fitting.'

V MEASUREING EQUIPMENT
Steel rule. Columbus gauge. Inside and outside calipers. Jenny calipers. Bell-punch. Surface plate—of plate glass for cheapness. Micrometer. Test dial indicator. Rule stand. Scribing block. Scriber. Engineer's blue or other marking bases. Measuring methods. Value of rough notepad. Do not trust dial markings—measure.

VI RANGE OF POSSIBILITIES
Universal nature of Unimat. Headstock mounting of work. Turning between centres. Centring work. Simple work between centres. Crankshaft turning. Use of collets. Use of height raiser for larger work. Hardwood block in lieu of height raiser. Grinding. Milling. Using tools with over-size shanks. Three-jaw for irregular shaped work. Balance of eccentricly placed work. Other work.

VII GETTING TO WORK
Horsepower of motor. Need for tool sharpness. Speed range. Speed and work diameters. Tool material. Belt slip to combat overload. Cutting speed tables. Lubrication. 'Suds' or modern improvements. Brushing it on. Tray for wet workers.

VIII USING THE TOOLS TO ADVANTAGE
Boxed tools offered as a set. Centre height requirement. Fix firmly. Cross slide action. Rake, wedge and clearance angles. Tools for different metals. High speed steel, carbon steel, tungsten carbide. Grinding down oversize tools. Use of grinding wheel. Goggles. Tool grinding jigs. Table for grinding angles. Boring tool. Moderate cuts. Trial and error. Parting-off. Support against sidethrust. Parting off with hacksaw blade. Taper turning. Holder for round section tools. Setting up for two sets of centres. False endpieces for eccentric turning. Flycutting. Boring bar. Value of vertical slide.

IX SPECIALIST ACCESSORIES
Screwcutting. Taps and dies in the lathe. Changewheels versus hob and chaser. Thread form needs. Metric or other standards. Thread cutting accessory is formidable. Thoughtful assembly. Important points to see. Test piece. Slow running speed. Collet chuck. Luxury collets. Make your own! Polishing spindle—and alternative. Watchmaker's collet spindle. Planing attachment and router woodworking only. Circular table and indexing attachment. Index gear table.

X THE LATHE AS A DRILLING & MILLING MACHINE
Fitting up the vertical column. Pinion racks the drill. Truly vertical need can be checked with faceplate. Faceplate as boring table or use the three-jaw chuck. Use of headstock raiser to extend reach. Machine vice to hold work. Adjust height to regulate reach. Milling. Slotting. Use up broken drills.

XI SIMPLE ACCESSORIES TO MAKE
    A contributor with ideas for the Unimat. Tape recorded instruction course. Scriber. Vertical slide. Drill/reamer stand. Tailpost dieholder. Slitting saw holder.

XII LATHE AS A WOODWORKING TOOL
    Limitations accepted. Ideal for small work. Pleasures of woodturning. Slide—leave in place or remove? Jigsaw and other accessories. Original pattern jig-saw. Current jigsaw and sabre saw. Larger saw table. Circular saw with and without table. Height raiser and larger diameter blades. Exact line cutting. Planer and router attachments. Sharpening cutters and routers.

XIII WOODTURNING ON THE UNIMAT
    Turning chisels. Prong chuck. Handrest location. Gouge and chisel Glasspaper for fine finish. Ideal woods. Contrast colour work. Plastics.

XIV WORKSHOP, GARAGE, KITCHEN TABLE . . or . .
    Locations and choice. Rust prevention. Workbench. Garage corner minimum size. Portable size. Portable bench for patio use. Cabinet to house lathe and serve as bench. Glass fronted wall cabinets for tools. Multi-drawer minicabinets. Materials. On/off foot switch. Chuckboard. Wooden blocks. Fixed steady of wood.

APPENDICES
    Technical specifications
    Accessories for SL and Mark 3
    Additional reading

FOREWORD ...

As a keen user and protagonist of the virtues of this little lathe since 1956 I was delighted to be invited to write a book about it and accepted with alacrity. I make no claims to be an 'expert'—it is a dreadful label to attach to anybody!—but I have been an enthusiastic user for over twenty years. I hope I can convey in print something of the very real pleasure and delight that my Unimats —yes! I have had five in all, with three in operation at the moment, including the very latest Mark 3—have given over the years. Each time I have 'grown up' to a bigger lathe and disposed of an old friend I have so felt the loss that I quickly acquired another. Present team comprises one fairly old of the 'middle period' of production, one nearly the latest apart from motor, one 'advance' example of the new Mark 3 ... and I nearly forgot there is yet another, the 'official' office Unimat at my place of work!

It is surprising that though more Unimats than any other make of lathe, bar none, have been sold in the postwar years so little has been written about them. Apart from the excellent operating instructions translated from the German text which come with the lathe sold in the United Kingdom, and a more profusely illustrated publication published by the original distributors in the United States, there has until recently been only very occasional—and sometimes condescending!—articles in modelling and craft periodicals. The Austrian manufacturers Maier & Co. have masterminded a tapebased instruction course for beginners, and a follow-up series for the medium skilled, projects which have attracted considerable support in Canada. I have not seen them advertised in England—certainly not brought forcibly to the attention of buyers. The past few months have seen the emergence of a real Unimat maestro who has been contributing advanced articles to the model press, and as this foreword is written we hear of a Swedish user who has incorporated a traditional set of change wheels and screwcutting facilities. I hope we are on the verge of a break through.

May I say thank you here to the various 'big lathe' operators who have corrected my divergent ways from time to time—from toolroom men from the car factory up the road with their gifts of highly unsuitable cast off tools and lots of good advice to 'Curly' Lawrence (LBSC) who tried very hard to wean me on to larger projects and referred for years to my 'mechanical mice' as he called my electric model racing cars ... and lots of others who remain anonymous but who have from time to time said 'why not do it this way ...'

I. THE UNIMAT LATHE

It can safely be said that the majority of the readers of this book will fall into one of three main classes. There will be the complete novice who has either just bought a Unimat, or is hovering on the brink of such a purchase; the rather more expert operator who has been using his for some time and is ready to indulge in a few accessories and extend his activities; and finally the dedicated enthusiast who is a compulsive reader of anything and everything about his chosen hobby tool. I hope we can find something to interest and help all these groups.

First of all we must ask the question why have a lathe at all? What can it do that cannot be purchased ready made, knocked up with hand tools or avoided by doing it some other way? Nowadays the Unimat is modestly described as a small machine tool, in earlier times as on the now dogeared operating instructions before me it was billed as a universal machine tool. It can in fact do all the things in skilled hands in miniature that would require an entire machine shop to handle in larger numbers and sizes.

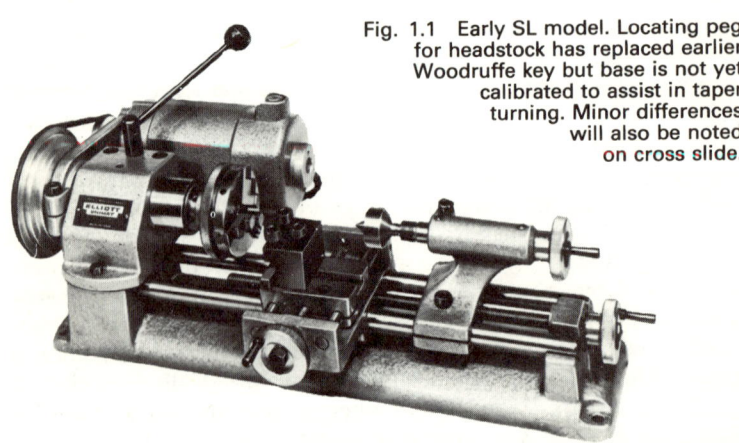

Fig. 1.1 Early SL model. Locating peg for headstock has replaced earlier Woodruffe key but base is not yet calibrated to assist in taper turning. Minor differences will also be noted on cross slide.

THE UNIMAT LATHE

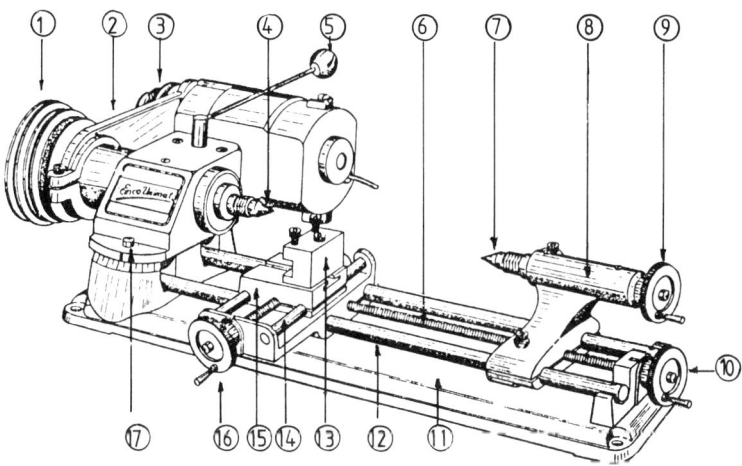

Fig. 1.2
1. Headstock Pulley
2. Motor bracket.
3. Motor pulley
4. Centre.
5. Feed pinion lever.
6. Lead screw.
7. Centre.
8. Tailstock.
9. Tailstock handwheel.
10. Handwheel.
11. Bed.
12. Slide bars or shears.
13. Toolholder.
14. Cross traverse slide bars.
15. Slide rest.
16. Cross feed handwheel.
17. Locator pin.

With the one machine it is possible to carry out a multiplicity of operations including turning (the first and original use of a lathe) milling, drilling, grinding, polishing, threading, jig and circular sawing in metals, plastics wood and many other materials. The only proviso it asks of its operator is that he or she treat it with care and respect and do not demand performances beyond its strength. For those with a desire or need to make practically any small item in virtually any material it will prove a remarkable tool.

There is no special mystique about the operation and functions of a lathe. Basically, it is a horizontal bed on which is mounted two pedestals, between which work is fixed and made to revolve against the cutting edge of a tool conveniently placed on a rest. This is all that many industrial lathes in fact comprise. But the small workshop man or amateur cannot afford either the space or the money to acquire a whole battery of specialist machines to do other engineering work, so this basic lathe conception has been cunningly elaborated to perform—with the aid in many cases of additional accessories—the formidable list of activities listed above. We still meet a lot of professionals who tend to sneer at the impossible versatility of the small lathe, but, if they can be persuaded to look and see for themselves can be convinced of its virtues. Indeed, in the now long

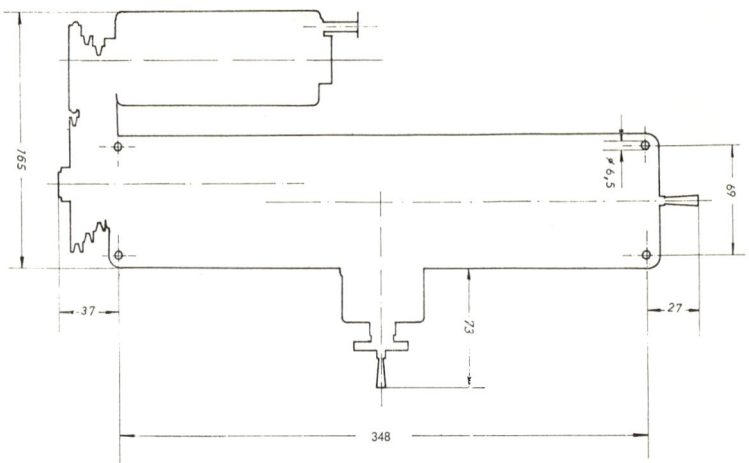

Fig. 1.4 Layout of SL Unimat, showing hole location for holding down bolts. Position on board is slightly towards front, leaving motor just clearly over board.

past war years, amateurs called to work in professional shops were often able to show their professional colleagues quicker and better ways of doing things to everyone's financial benefit and the extreme chagrin of that traditional machine tool operator's enemy the ratefixer!

The Unimat first came on the market over twenty years ago, and is today basically the same instrument. A few changes have been made, the original 40 watt electric motor on my first lathe, gave place to a 75 watt motor, and the fourth and latest of my collection now boasts a 90 watt motor. There have been minor changes to the design of the headstock casting—the addition of divisions to enable a measured amount of taper offset; a different type of tension screw, some cosmetic changes to the tailstock and other minor amendments.

But we must now describe the lathe and would urge the newcomer to study the named parts on the explanatory diagram so that future reference to them will be quite clear. The design is unique in that it contains a self-contained headstock unit with motor and countershaft. This is usually mounted on the bed, but can be raised vertically and attached to a separate pillar provided when it operates as a sensitive power drill. Instead of the more conventional cast shears the saddle, tailstock and cross-slide move on slide bars. This is quite common practice for small lathes—it has the advantage that swarf tends to drop right through to the main bed and can be cleared away easily. Against this, the design prevents the incorporation of a

THE UNIMAT LATHE

gap which could offer room for larger diameter work to be swung, though not recommended in a lathe of this modest size. As will be seen, even this apparent disadvantage is overcome with the accessory raising piece which gives an extra ¾ in. clearance.

The headstock is fitted with a cylindrical sleeve in which the threaded headstock spindle is housed. The whole sleeve can be moved sideways with a rack and pinion movement, or can be locked solid. The threaded spindle has a cone pulley attached to it to carry a V-belt. The motor bracket in standard form allows for a motor pulley and an idler pulley. These with the main pulley can all be removed and reversed and permit a range of some eleven speeds. The substitution of an accessory motor bracket with provision for an additional idler pulley enables two further low speeds (useful with ferrous metals) to be used. The belts provided are of rubber and will naturally have a limited life. This can be prolonged by pre-stretching before use, lubricating with a little glycerine (not too much—it will cause slip!) and keeping the lathe from direct strong sunlight. If work is unlikely for a period remove belts, dust with boracic powder (baby powder) and store in a screwtop jar. Nevertheless, keep a spare set handy—they always go on a Sunday!

How you receive your lathe may depend on where you live! My very first one, as befitted an instrument from wood-oriented

Fig. 1.3 A later model SL with modified cross slide casting, calibrated headstock base for taper turning. Self act has also been added. Original type alloy handwheels have been fitted from an earlier model for purely cosmetic reasons. Standard handwheels are now black plastic as fitted to Mark 3.

Austria came in a beautiful polished wooden box, intended to retain it and some accessories during its working life. Actually it was a bit awkward for regular use, and a little too small when the lathe was mounted. The second lathe came secondhand so its original packing state is unknown. Number 3 was in a plain cardboard box in a colour matching the lathe finish. Latest, Number 4 was the best packed from handling point of view in stout card, with expanded polystyrene mouldings to take the various parts.

The motor bracket will have to be fitted to the headstock and motor and the pulleys assembled. This is very clearly set out in the assembly instructions. A plug must be fitted to motor lead, which is already provided with an on/off switch, originally in lead, now on motor.

Where and how it is to be used must decide the next steps. Originally my lathe was screwed to an old ironing board, just long enough to go across the arms of my fireside armchair! The winter was cold and this proved a splendid arrangement, apart from the swarf down the back of the armchair. Today my lathes are mounted on smaller wooden bases, with handgrips each end as illustrated. They can then be used either on top of my tool cabinet and lathe cupboard (or stored inside it when I want my workroom to look specially tidy); or can be taken into the garden and located on top of my Workmate portable bench whilst I sit beside it in the shade and enjoy my work happening.

I have tried both painting the base with white enamel paint and the alternative of facing it with Formica, which looks very pretty.

Fig. 1.5 Simple mounting board for either Unimat model. Length about 18–20 ins., width 8–11 ins. Thickness of main board not less than ¾ in., similar thickness of cross pieces. Note indents to enable lathe and board to be easily lifted. Paint or cover with Formica.

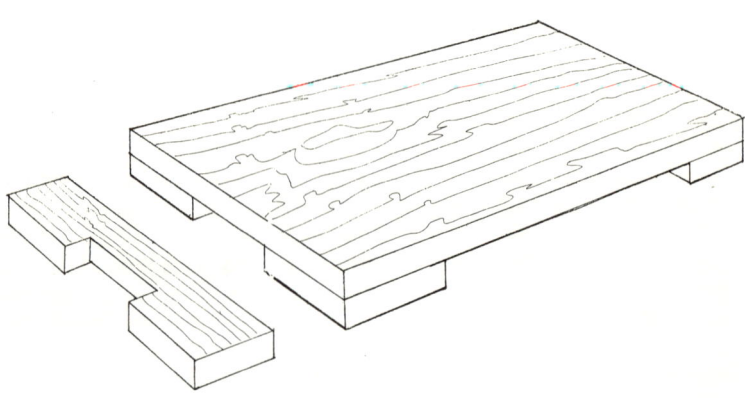

THE UNIMAT LATHE 15

Both styles clean up easily—the paint can always have another coat. On balance I prefer the paint job and will probably strip off the Formica eventually.

Well, there it is, simply asking to be used for something! In minimum basic form you will also have a faceplate, a lathe dog, two centres, a slotted screw, grinding wheel holder, tool-holder, vertical column and bracket. There is a locking handle to move the headstock sleeve and an allen key to tighten sundry allen screws. As a precaution slap a bit of bright paint on the allen key—it gets mislaid very easily. Not supplied but necessary if the main belt pulley needs to be reversed is a good spanner of 19 mm across faces (a/f) Do not economise in the quality of this spanner—a bad fit and the hexagon retaining nut will lose its corners and be increasingly troublesome to shift.

Three handwheels will be noted. One serves the tailstock and moves the tailstock sleeve in and out; another works the cross-slide and the third drives the cross-slide bed lengthwise on the long threaded spindle. All these handles will be seen to be marked with divisions. They are not very clear and it is a good scheme whilst everything is fresh and new to fill them with contrasting paint, wiping off the surplus and allowing to dry. Each division marks a part of a turn amounting to .05 mm. There are twenty such divisions which means that a full turn of the wheel represents a movement of exactly 1 mm. But remember that by cutting off a slice all round a bar you have in fact reduced its diameter by *twice* the indicated amount of cut! Measure a test piece in due course and see for yourself if this is not quite clear to a beginner.

If you are happier with inches then, in round figures each mark equals .002 in. However, do not necessarily get into the habit of regarding these figures as exact, since there is some amount of play or backlash. This can be adjusted by tightening the nut on the handwheel but some play will exist and precision will come better by frequent measurement rather than complete trust in the markings.

On the cross-slide reposes the tool-holder. As well as the crosswise movement of the cross-slide and the lengthwise movement of the leadscrew handwheel, the tool-holder can also enjoy a limited travel along the single T-slot on which it rests and can rotate through 360° thereon. A tool fitted in its slot can be adjusted almost infinitely on its plane. A limited amount of height regulation is also possible if the tool is packed up underneath.

The tailstock is located on the right and can be moved along the slide bars in either direction. It is not geared into the leadscrew and when located as needed is fixed by tightening its allen screw. The tailstock barrel can be advanced or retracted by means of its handwheel. In the barrel the centre can be placed for turning between

16 THE BOOK OF THE UNIMAT

Fig. 1.6 Young man at work has excessively tidy bench! This posed picture shows bar stock being turned down in steps, using ball bearing centre and three-jaw chuck. Lathe is not mounted and lead has been tucked under base for easy access to on/off switch.

centres, or, if you have sported one, the ball bearing live centre goes here. A drill chuck can be screwed on the barrel or can be screwed on to the headstock main spindle.

The faceplate can also be screwed on here to hold work bolted on to it, useful in modulation for milling set-ups.

Finally, if we do not want to use all this as a lathe we can remove the headstock and motor, insert the vertical column and headstock adapter and we have a sensitive drill set-up as soon as we have screwed on the chuck.

Other useful bits and pieces we have not specifically mentioned include the grinding wheel adapter and the locking handle—this is particularly useful when drilling.

When you re-assemble the unit as a lathe you will find out how useful is the little pin which fits just below the nameplate. This headstock alignment pin joins up the headstock and the main lathe bed casting to ensure that it sits exactly parallel with the slide bars.

Upkeep & Maintenance

An old adage with things mechanical was: 'If it's working well

THE UNIMAT LATHE

—leave it alone!" Manufacturers of the Unimat seem to share this opinion. In my oldest set of instructions advice is to re-grease headstock spindle bearings every 300–500 working hours; my latest has advanced this to 1,000 hours! It is not a major job but requires a 19 mm spanner to remove the large nut holding the pulley on. You should have one standing by anyway, it is rather slovenly to rely on an adjustable spanner. To quote the earlier manual, after removing belt pulley, feed lever and headstock bolts and set screw the spindle can be slid out towards the tailstock. To protect the spindle thread screw on the faceplate first.

Wash out the two ball-bearing races in paraffin or petrol—putting them securely in an old tobacco tin or similar and ensuring no balls are allowed to stray. It is a wise precaution to have a few spare balls in case any are damaged or lost—they should not be difficult to obtain from your stockist—or from garage spares department. Regrease, using a named brand high melting point grease, and re assemble.

Just how long it takes to build up 1,000 working hours is a matter for you. At ten hours a week (2 hours for 5 evenings) it will take a year to clock up 500 or two years for the thousand, so it is not likely to be a frequent demand on your time. More often must be the day to day maintenance of the machine. It is sound policy never to put it

Fig. 1.8 Martin Evans, well known writer on model engineering and live steam and for many years editor of *Model Engineer* turning a 3½ in. gauge locomotive wheel on the Mark 3 Unimat in the comfort of his office.

away, or stop work for the day, leaving it swarf coated and damp with suds or the like.

Keep an old paintbrush—about ½ in. to 1 in wide—as the regular brushing down tool. Add to this a bottle brush of medium size (your door-to-door brush man will probably give you one as a makeweight!) which hangs up right by the lathe. This is to brush out threads of chuck, faceplate, drill chuck, tailstock and spindle so that no chips are imbedded in them. Add light machine oil and wipe over everything with a rag. Always leave chucks with threads uppermost, and use the chuck board when removing from spindle. Special attention should be given to cleaning up after a grinding session—you have grits as well as chips to cope with.

Wear and tear on chuck jaws should not be too heavy unless you actually make a habit of dropping them. Handwheels may work loose from time to time but can easily be tightened up—but not so tight as to make their operation tiringly stiff. Very few parts of my oldest Unimat have been replaced in twelve years of weekly at least if not daily use.

My old friend Curly Lawrence—one of the fastest of workers—used to keep a dustpan and brush right by the lathe and clear up after every operation. But then he kept the tidiest workshop ever seen!

II. ADDITIONAL EQUIPMENT AVAILABLE

When it comes to lathe accessories the pundits—whoever they may be—will discuss unendingly just what is essential to the well equipped operator. In due course we will describe the use and abuse of the principal items, and now for a start give our own short list.

I think we must assume that the lathe has come equipped with the items enumerated in Chapter I, including the three jaw chuck duly matched to the lathe. This gives us discussion point Number 1! Why a three-jaw? Would it not be better to have the four-jaw chuck which will accept workpieces of all shapes instead of the more limited acceptance of the three-jaw; and the price difference is so trifling. Perhaps we have been hypnotised into believing that first need is three-jaw. I have erred in this direction; but am rapidly being convinced otherwise. Snag is that the latter tool is rather more trouble to adjust correctly—but what lovely practice!

Then we simply must have something to hold things firm either on the cross-slide for the lathe or when used as drill on the drill table, so our next essential is the machine vice. There is really little controversy here, it is a worthy second need and will earn its keep in no time.

What next? A school of thought will be loud in lobbying for their favourite, claiming that as *turning* is so vital a part of lathe operations, and turning between centres is what they really mean, then a nice double ballbearing live centre should be the choice. But wait, the lathe comes with two dead centres surely these will suffice for early days? Rather reluctantly we concede the point and move on to the next claimant—a drill chuck for mounting on the tailstock, or for use on the vertical column. Making holes in things is going to be a major work so we hope you will agree with this need.

We must also accept that the operator has acquired some lathe tools which we will cover in the appropriate chapter. For the rest we have devised two 'league tables' where we list other accessories in what we consider order of importance. Two lists, one for the metal work oriented user and one for the woodworker. Do not please try to get them all at once unless you have a lot of spare cash or have won a prize. It is far better to acquire them as the need arises and the nature of what is to be your main interest crystalises. Then you can make known your needs to the family and there need be no doubt at birthdays and Christmas that a tool for the 'mangle' will be accepted with joy . . . half my etceteras came that way and give me special pleasure to use.

The three-jaw self-centring chuck is likely to prove a most useful accessory. In the early days of manufacture it was customary to include such an accessory in the 'package deal' and most retailers will urge the buyer to do so today. It is desirable that the chuck is mated accurately with its flange, and, preferably fitted to the lathe on which it is to be used. However, in over twenty years of use and in assisting in the purchase of many a lathe for friends I have never failed to find that the makers have completed the job already! Improved methods of manufacture now enable chucks to fit without individual attention.

If you should find signs of non-concentricity in yours then is the time to look up the instruction manual and follow the steps given.

Fig. 2.1 Belt guard a newly introduced accessory for Unimat SL. With latest on/off switch located on motor this is a desirable additional accessory. A single butterfly nut holds the retaining screw so removal for belt change is quick and easy.

ADDITIONAL EQUIPMENT AVAILABLE

Once upon a time it was usual to provide two sets of jaws with a s-c chuck, one set serving to hold work on their sharp points for drilling, and other being stepped to hold larger pieces of work. Now both uses are combined in the single set of jaws. These jaws are machined to fit into T-slots, and have teeth on their rear faces to engage with machined scroll that winds in like snail shell's markings. It is therefore important when jaws have been screwed right out for cleaning or reversing that they are re-inserted in the order of their numbers 1, 2, 3 which are engraved both on the jaw and on the body of the chuck. In the Unimat chuck the scroll is operated by knurled external ring, with two holes therein for the insertion of two tommy bars provided for that purpose. Do not tighten too hard or it may be difficult to unscrew, damage the work, or injure the chuck threads.

To re-insert with the pointed or drill end inwards, order is a straightforward 1, 2, 3. Push in No. 1 at the beginning of the spiral thread until it is just gripped, twist round clockwise until the thread beginning is similarly located for No. 2, then on to No. 3. Twist the ring until the three jaw points meet in the middle. If they do not—then you have done it wrong! Unscrew and start again. It is quite finicky until you get the knack.

But wait! If you are using the jaws the other way round, that is to make use of the steps to hold some larger piece of work, re-insertion is NOT 1, 2, 3. Instead the reversed No. 3 goes into slot 1, then 2 goes in slot 2 and finally 1 into 3.

Remember that not only can the stepped parts of the jaws hold the outside of a workpiece of ample size when they have been reversed. It is also possible to hold a hollow workpiece with jaws in unreversed positions. Here, of course the knurled ring must be opened wider to get a firm grip instead of closed.

Although normally for round or hexagonal workpieces, square or rectangular work can be held if first made round. It is also possible to make use of the jaws to hold irregular work on occasion—though not a course to be recommended for regular operations.

For these irregular shaped pieces the four-jaw chuck is normally used. Here the four jaws, also numbered are separately tightened, so that they can occupy any position in relation to each other. Locating work in the precise position required for the operation in hand requires considerable care. It is helpful after fixing approximate location to insert a centre in tailstock and bring it up just to touch the workpiece when final adjustments can easily be made.

Accurate setting-up comes from experience, but there is one simple way that embodies no special appliances other than a piece of chalk. Run the workpiece in the chuck fairly slowly and just touch it with a snub nosed piece of chalk hand-held and supported by resting on the tool post. The amount of chalk deposited will show

Fig. 2.2 Three and four-jaw chucks. Three-jaw has jaws reversed to hold larger workpiece; four-jaw has its jaws in normal fashion to take small items.

the high spots, and the jaws can be loosened and tightened until a true setting is obtained.

Too much care cannot be taken of the chucks—their long life and continuing precision depend on careful handling so that a simple chuck board or wooden platform to go under the headstock to receive them as they are unscrewed will avoid danger of metal dropping on metal damaging either the chuck or the shears. They should be kept adequately cleaned—retaining an old $\frac{1}{2}$ in or $\frac{3}{4}$ in. paintbrush for cleaning swarf out of the ways and a regular oiling after use. When put away they should always be laid with the jaws face down and the threaded flange uppermost to protect its precious threads from harm.

Note that the four-jaw can only be used at its fullest extent with jaws reversed if the accessory intermediate piece is inserted between headstock and bed. This raises it approximately $\frac{3}{4}$ in. and allows clearance. However, since we are making full use of this part in other ways its use will occasion no surprise.

The automatic feed attachment, or self-act, may lightly be regarded as a luxury if it is considered merely as saving the operator the tedium of feeding long cuts by hand. This is far from the main purpose, useful as it is, of the device. First we should consider exactly what it does. It is an accessory of the Unimat (*though now built in as an integral part of the latest Mark 3 model!) that gears the leadscrew to the mandrel in such a way that the lathe carriage, or longitudinal slide, is made to advance the tool along the work

*Not sold as part of the package: charged extra if required.

ADDITIONAL EQUIPMENT AVAILABLE

automatically. Its speed of advance will be related to the operating speed of the lathe at the time since it works directly from the last pulley actually driving the mandrel.

Since its speed is constant—far better than could be judged operating the handwheel—which ensures a better finish for the work it should always be of value for at least the finishing cut.

It is one of the more recent additions to the range of Unimat accessories but should be fairly high on the list of priorities. Assembly is easy. The main part is a base plate, or rather plates, joined by the feed shaft. At one end is the pulley which connects to the spindle head and at the other a spring loaded worm wheel which operates a toothed handwheel replacing the standard handwheel. The lathe bed is unscrewed from its base and refitted on top of the baseplates by means of four hexagon headed screws. These are fitted from above into the threaded holes in the baseplates designed to receive them (not as shown in the Unimat Instruction manual from underneath the bed!). Final job is to remove the hexagon nut from the large pulley shaft and replace it with a slightly large nut of same thread diameter which contains at its end a further pulley to drive the feed shaft. A 19 mm a/f spanner fits the standard retaining nut; the replacement large one requires a 22 mm a/f size, or a good stout adjustable spanner will be suitable.

Attach rubber belt supplied, lower spring loaded worm wheel on to toothed handwheel now fitted by means of the little knobbed lever provided. Check that feed shaft is running freely, add a drop of machine oil to each lubricating point on the bearings, and switch on. It will be seen how slowly the carriage advances (less than one-thou.

Fig. 2.3 Clamping plate, offered as part of indexing accessory but useful too as stout faceplate with T-slots and three threaded holes, plus concentric centring grooves. On right is standard alloy faceplate—a much lighter device. In between bell-punch for bar stock.

of an inch per revolution) and this may well serve as a guide for the ideal speed of a cut. Travel is towards headstock; by crossing the belt into a figure eight direction is reversed for cuts towards the tailstock. When not required drive belt should be removed though the basic equipment should be left permanently in place.

ACCESSORY 'LEAGUE TABLE'

Item	Metalworker	Woodworker
Four-jaw chuck	1=	
Three-jaw chuck	1=	
Machine vice	3	
Drill chuck	4	3
Ballbearing live centre	5	
Swivel toolrest		1
Spur drive centre		2
Slow speed attachment	6	
Selt-act	7	
Milling table & clamps	8	
Thread cutting attachment	9	
Jigsaw & blades		4
Circular saw attachment		5
Increase height adaptor	10	6
Mitre gauge	11	7
Indexing & 48 div. plate	12	
Circular table	13	
Sanding plate		8
Grinding wheel guard	14	9
Planing attachment		10
Routing attachment		11
Fixed steady	15	
Truing diamond	16	12
Rubber disc with arbor		13
Polishing spindle	17	
Collet attachment	18	

Above list assumes appropriate metal and woodturning tools as 'musts' already. Significantly it does *not* include woodworking conversion set (longer shears for bed) since if woodworking that important next stage is woodworking lathe! (Sorry Mr. Unimat!) Flexible shaft and watchmaker's collet sleeve omitted for very similar reasons that specialist equipment of this sort is over-taxing

ADDITIONAL EQUIPMENT AVAILABLE 25

Fig. 2.5 Round work held in machine vice for vertical drilling of hole for tommy bar on tailstock dieholder.

the Unimat if extensively used—don't be mean get the tool for the job!

Any favourites list is open to discussion—even to argument—so protagonists of any particular tool or accessory are invited to state their case!

Some additional needs will certainly arise when the more interesting of the Mark 3 accessories have been in use long enough to assess their special value.

Fig. 2.4 The very beautiful collet holder with a selection of collets and threaded headstock attachment piece.

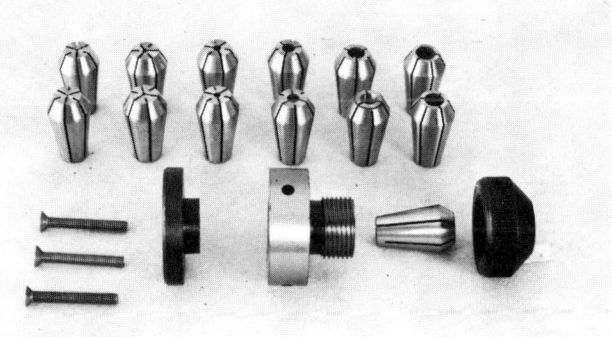

III. THE NEW UNIMAT MARK 3

There comes a time in the production life of the most successful tool that detail improvements have reached their effective limit and the harsh decision must be taken to launch a radically new model. For some twenty years the Unimat SL has been the subject of minor improvements, though the original conception was such that these changes needed only to be minor. Parallel with these changes came a wealth of accessories such that the needs of a great number of different groups of users could be satisfied.

Now the step has been taken and the Unimat Mk. 3 has been released. It follows very closely the pattern of the 'big brothers' in the Emco range—the Emco Maximat and the Emco Mentor—and yet retains the most worthwhile characteristics of the original Model SL. First of all size has been slightly increased, notably in mandrel

Fig. 3.1 Unimat Mark 3 in its black and white livery. Note black plastic handwheels. In this picture self-act which is an extra has been fitted, and graduated compound slide which is also extra is installed.

THE NEW UNIMAT MARK 3

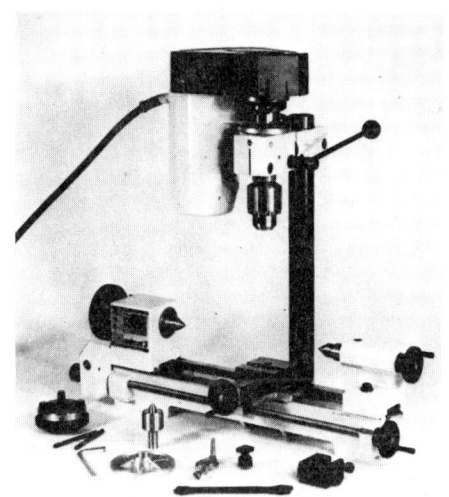

Fig. 3.2 Here the vertical drilling spindle has been erected, where it may safely be left permanently. Motor assembly has been removed from headstock and locked in place on drilling spindle. Accessories shown include three-jaw chuck, sundry keys, faceplate, ballbearing centre, dog, spanner, threaded T-nut to hold faceplate/chuck on cross slide, standard toolpost. Drill chuck is in place.

bore and diameter and in the capacity of the drill chuck, now up desirably to $\frac{5}{16}$ in. capacity instead of $\frac{1}{4}$ in. But most important change is undoubtedly to the basic design. The old round shears that bolted down on a base have been replaced by a 'large lathe' bed of cast iron in the favourite H section with the top surface machined to carry the cross slide and tailstock.

Headstock is now fixed, though made as a separate piece, with an ingenious steel plate carrying the motor and drive pulleys. Vertical

Fig. 3.3 Mark 3 layout. Note that only *two* fixing screws required for attachment to baseboard.

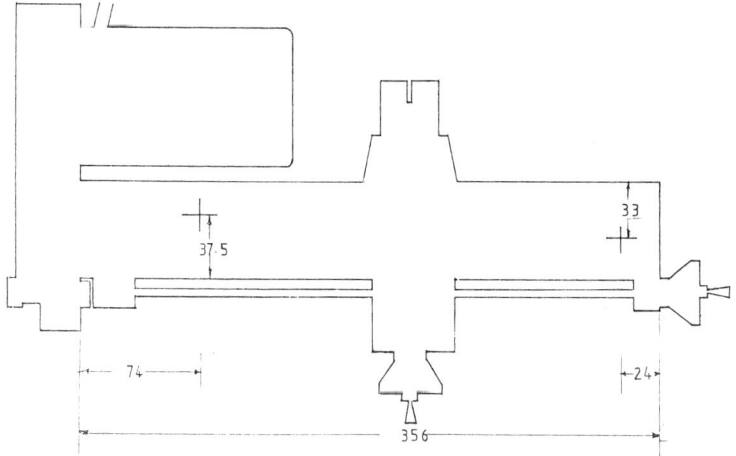

Fig. 0.4 Clever detachable motor assembly with pulleys and belt cover. This can be removed from headstock and put on drill column, or vice versa in less than one minute on the stop watch! Less if you really hurry!

spindle for drilling and milling is fixed in place at the centre of the lathe (here the resemblance to the larger lathes in the range is particularly noticeable). The steel plate carrying the motor and drive pulleys has a quick change bayonet fitting and can be swiftly unlocked by loosening two hexagon bolts and placed in position for drilling. This change takes less than one minute without trying to hurry!

Motor power has been stepped up slightly to 95 watts and on/off switch which was originally found along the flex and more recently on the motor now operates a two-speed rocker. In the general tidying up of this section a neat hinged protective cover has been fitted to go over the pulleys, thus obviating any risk of touching a moving rubber belt when switching on or off.

Motor pulley unit has identical diameter grooves each side of a larger groove, providing two size options and two locations to reduce friction by asymetric banding. Pulley on machine has two sizes of groove (in lieu of three on Model SL) Idler pulley has the usual three, stepped down in the approximate ratio 4: 2: 1. Speed changes can all be effected by band changes without removing or turning round the pulleys: another speeding up and simplifying process. This gives eight speeds 130—200—350—560—920—1500—2450—4000, achieved with two motor speeds and two optional pulley sizes.

THE NEW UNIMAT MARK 3

Fig. 3.5 Head of vertical drill column showing sprung quill and graduated dial enabling exact set ups to be repeated. Fine feed attachment screws on here as additional accessory.

Fig. 3.6 Close up of standard tool post which indicates clearly V-section of slide and gib.

Fig. 3.10 Man at work! Bench light is a great help in doing fine work. Seated position (see also other working pictures) is recommended for long sessions in complete comfort.

Fig. 3.7 Tailstock end of lathe bed to show robust nature of solid bed and stouter build of tailstock casting.

THE NEW UNIMAT MARK 3

With a conventional lathe bed it is now possible to bring the leadscrew from between the round shears to the front of the bed. Not only does this tidy up the design by increasing the gap between the handles of leadscrew and tailstock but also enables a much neater and less cumbersome self-act to be fitted as an extra.

Cross slide has been re-designed with prismatic guides and gib with adjusting strip. This has a far more robust feel and is clearly 'man enough' to tackle quite formidable jobs.

Minor points include TV suppression, quieter running and an increase in speed range (130 to 4,000 rpm) plus quicker speed alterations.

A neat little plate appears on the headstock setting out the various speeds, eight in all, which can be obtained very simply with the two motor speeds. It will be noted that the low 130 rpm which needed a special motor bracket and an additional pulley is not required. This in itself reduces the friction losses of power and brings more to the actual work of what has left the motor.

Some considerable improvements have been made to the vertical column when used as a drilling/milling machine. The familiar locking handle that served both horizontal and vertical uses is now of course only required for the drilling machine. An added coiled spring enables this use to be improved when employed as a quill. The head which used to be adjustable on the SL retains this feature

Fig. 3.8 Speed/belt chart is in form of neat plate attached to headstock. Note also crossed belting that operates self-act, which is engaged by knurled knob front left.

Fig. 3.9 Compound cross slide with 30° marking range against location mark on base. This should prove an invaluable accessory and indeed is essential for taper turning on new model.

but has a graduated scale added which should be of particular value in drilling at a specified angle. It will also be noted that the top of the column is also threaded and this can be screwed on to the bridging piece or other accessories with or without the motorised head.

Improvements are also envisaged for the screwcutting hob. This will now be threaded on to the outside (i.e. through the main pulley of the headstock) and a bar will then swing over the headstock and connect up with the bed. This will enable longer screwcutting runs and of course follow very much larger industrial procedure. Mandrel is hollow and long rod up to a maximum of 6 mm diameter can be fed through. With a fixed head the useful option of turning the mandrel to allow taper turning is lost. Happily it means a reversion to more standard methods without risk of losing accuracy on the bed. An additional accessory in the form of a topslide with graduated scale will enable accurate taper turning to be carried out.

The increase in size of mandrel and tailstock diameters means that a number of accessories which screw on to the SL will be too small for Mk. 3. This is an unavoidable result of the improved specification. As will be noted when discussing the new range of accessories opportunity has been taken to build in a number of improvements that go beyond cosmetic alterations.

IV. MARK 3 ACCESSORIES

Far from being a mere cosmetic improvement it is very certain that the new Unimat Mk. 3 is rapidly approaching 'toolmaker' standard by virtue of the splendid range of accessories available. There may be a natural regret that so many of the SL Unimat accessories are incompatible but this is the price of progress. Even such items as T-nuts have been re-designed for Mk. 3 in a more robust form.

We have mentioned the normal expected minimum of equipment needed to operate the lathe. This includes the very useful 3-jaw self centring chuck now made with flange as an integral part which obviates any need to true up the parts as sometimes required with the earlier design. It also provides a stouter and more robust piece of equipment. The 4-jaw chuck has of course always been made in this way, and follows the same pattern with the larger size of mandrel fitting.

Fig. 4.1 Self-act gear box, engaged by pressed knurled handle and turning when it connects with dog on self act screw. Drive is, as before by the main drive spindle on pulley.

A much improved face plate is also available with four tee-slots and clamping dogs this is indeed a great advance from the light alloy faceplate originally offered with the SL. This is by no means intended to disparage the earlier model but to point out the very practical advances made with the new!

Perhaps the most appreciated all-purpose newcomer in the accessory field must be the compound slide. This is an additional part which fits on the slide carriage with its own toolpost. The slide carriage has two indicator marks on its surface and the compound slide has a 45°–0°–30° dial enabling a 75° range of settings to be made. Primarily this is intended to enable taper turning to be carried out on the lathe in spite of the headstock now being fixed in position. Now both outer and inner taper turning to a maximum length of 40 mm ($1\frac{9}{16}$ in.) is possible. This means that the delightful little col-

Fig. 4.2 Fine feed attachment to drilling column which completes usefulness of this much improved accessory.

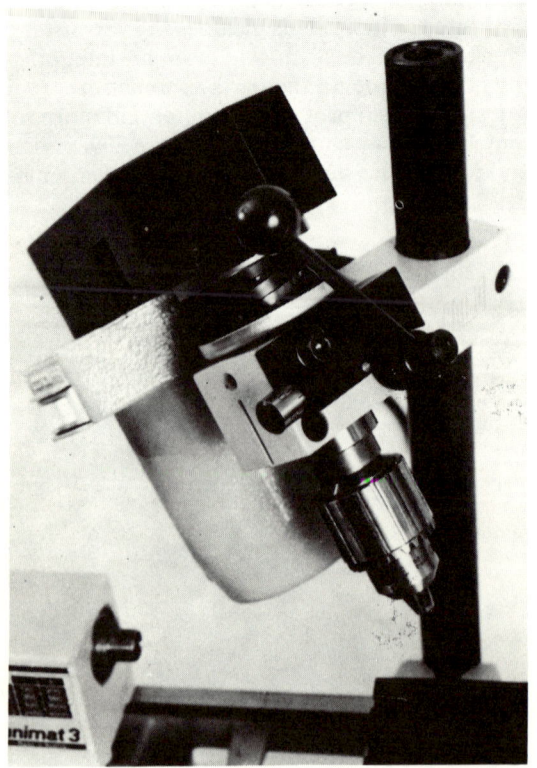

MARK 3 ACCESSORIES 35

Fig. 4.3 Another location for drill column, which has one end threaded to enable it to locate via T-nut with this bridging piece. It can also be attached direct to cross slide or to milling table.

Fig. 4.4 Here the compound slide is set up for taper turning. Maximum length of taper is 40 mm (about 1⅝ in.)

Fig. 4.5 A nice piece of work! Inside taper turning using inside turning tool and with compound slide set at 30°.

lets available as an accessory can be made on the lathe if the impecunious owner so desires! There are hosts of other uses for this set-up as it provides complete and repeatable 360° movement in a refined manner.

Introduction of fine feed to the vertical drill is another valuable feature. Until now even with the springloaded feed handle operation has required some manual dexterity. The fine feed reduces this to a matter of setting only and enhances the already useful indexed head. It seems likely that considerably greater use will now be made of the vertical drill.

This comes as an additional part (not included in the original package as before) and is left permanently in place. The drive unit is listed now as a separate unit and many users will wish to have this separate motor and intermediate gear on the carrier plate as an extra. Although it takes only a minute to change motors there are occasions when it will be desirable to leave lathe motor in place.

Provision of a flycutter will also enhance the uses of the vertical column and the versatility of the lathe. This is a fairly simple accessory and many users will undoubtedly have already devised their own version.

There are times, however, when the 'permanent' position of the vertical column does not apply. The head of the column it will be noticed is threaded to take a T-bolt. This can normally be left in

MARK 3 ACCESSORIES

place there until wanted. The column is then removed from its upright position by unscrewing the two bolts holding it to the lathe bed; reversed so that the top becomes the bottom, when it can be slid on to any of the slots on the milling table. A special mounting bridge is also available to take the column and thus increase the range of work possible when milling between centres. This mounting bridge slips onto the cross slide in place of the toolpost.

To continue the versatile uses of the drill column it can also be used to carry a slitting saw in the cutter arbor accessory. A similar fitting is described amongst accessories for the SL to be made on the lathe. Should you be thinking of making up your own collets a considerable amount of slitting will be required, whether you make your own simplified pattern or the more elegant production style.

Again the column can be mounted direct on to the cross slide without the intermediary of the bridging piece.

A very special feature of the SL has always been the threading device or hob. It is a thing to love or hate and many may be put off by the elaborate fitting instructions. Once in use and with a little practice it will be really appreciated and can be fitted or removed in very quick time. Mark 3 has refined it much further with the leader now fitted on the *outside* of the bed to the left of the driving pulley where

Fig. 4.6 Milling a flat on bolt head, using milling cutter in collet holder, with vertical column set on bridging piece. Note also key in fine feed attachment to provide progressive cuts.

it can be screwed in, a convenient hole in the belt cover permitting it. Apart from increased strength the device now follows very much the pattern of the smaller SL layout. It clearly enables greater lengths of work to be screw cut. The usual range of leader/follower sizes are available, namely nine for metric sizes from 0.5/1.5 mm pitches, and fourteen for inch pitches 16/56 tpi).

Indexing and dividing attachment again figures in the list, with 30, 36, 40 division plates as extras, a new size for 24 divisions going with the attachment in place of the originally included 48 divisions plate—a change which in no way reduces its usefulness (i.e. 3×2×2×2).

Three point fixed steady follows the general style of the SL accessory but is styled more elegantly to suit the black and white Mk. 3 finish. Similarly the grinding wheel guard is more robust and stylish than with the SL, and is bolted down on to lathe bed. It is also provided with tool rest and a chip guard. May we repeat—do cover up the lathe bed when using it with a cloth or cover of some sort. It is more than ever important with a flat bed to avoid grit thereon much of which would fail to adhere to round shears!

The new three-jaw chuck is a much stouter job than before, being made with backplate integral in the style of the original four jaw. Faceplate continues to be of alloy.

Fig. 4.7 Collet holder used to hold bar material for turning. Standard tool post is in use, and tool has minimum projection from holder for rigidity.

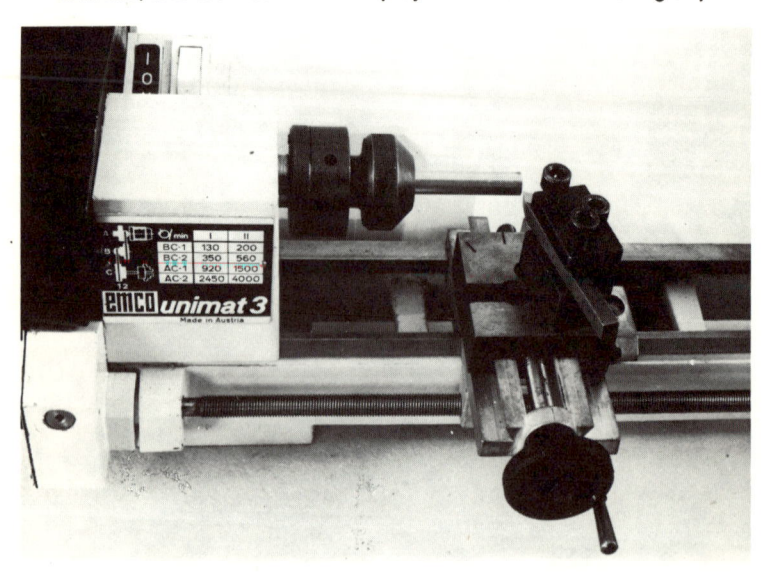

MARK 3 ACCESSORIES

Fig. 4.8 Threading attachment for Mark 3. Leader goes on *outside* of headstock spindle, enabling quite generous lengths to threaded between centres. Follower is much more robust than before, with improved threaded guide which rests on cross slide. It is no longer necessary to attach leader to chuck.

By bringing the leadscrew to the front of the bed it has been possible to offer a simple and compact self-act accessory. Indeed it seems so very much part of the design that in an initial review of the Mark 3 model I failed to appreciate it was an extra at all! A single hexagon cap head screw secures it to the side of the headstock, a slotted fork connects with the end of the leadscrew or can be simply pulled back out of contact. A rubber loop in figure eight form goes behind the main drive pulley on the headstock to fit in an already provided groove. The necessary worm gearing of the device is entirely concealed in the neat fitting styled to match the rest of the lathe, with instruction plate on its side. Untwist the figure eight loop to reverse direction as with the earlier more cumbersome version on the SL (In spite of its finger catching awkwardness it can come in very useful on the old favourite—I would not be without it now!).

V. MEASURING EQUIPMENT

Measuring equipment and marking out go together like fish and chips or bacon and eggs. Without the former it is hard to indicate work to be done and then finally the measuring equipment comes back to check that work has been done as required. The degree of accuracy required will to a large extent determine the quantity and quality of measuring equipment needed. Perhaps I should add here that a great deal *can* be accomplished with little more than a steel rule and a scriber but things are much easier with a little more than that. Again, quality tools will last a lifetime—so obtain the essentials in the very best quality that you can afford. One good way to get such items as micrometers, test dial indicators and the like is to belong to a model engineering club where members will often pass on excellent material at a low price when they are moving on to bigger and better; or there may be a club sale—auction or jumble!—with bargains galore!

Fig. 5.1   The very rudiments! Steel rule, homemade scriber and small square with adjustable arm. Cheap but indispensible.

MEASURING EQUIPMENT

Fig. 5.2 Moderately priced six-inch Columbus vernier gauge, which enables inside and outside measurements to be taken as well as use as a depth gauge, via small protruding probe on right.

First requirement is clearly a steel rule. With metric looming though not as yet generally accepted one with both inch and metric scales is desirable. One foot long certainly, plus at least one six-inch for reasons which will appear later. A sliding caliper gauge, also known as a Columbus gauge or Vernier gauge is the next requirement. This will normally enable measurements to one thousandth part of an inch, or equally as accurately in millimetres. They permit both inside and outside measurements and also have a depth gauge. An instrument giving up to about eight inches should be large enough. I also have a giant vernier nearly thirty inches long—it was a present, is very handsome but I have never used it in earnest!

Then we should get inside and outside calipers. They are available as both plain calipers, that is with a simple hinge, or spring bow calipers where opening is controlled by a threaded rod against a spring. These latter are worth the extra. Various sizes can be seen—those offering about a three-inch maximum opening should do very well.

Another useful variant of the calipers is the odd-leg calipers commonly called a Jenny Calipers. This is most useful for centre finding on the end of a round bar or locating points in relation to edges and in many other ways. A bell-punch is also of great value in centre finding particularly with diameters rather too small for using the Jenny.

Then a surface plate on which to rest work so that a flat reference can be obtained is useful. A cast iron plate is very expensive and need not be considered. Visit the local glazier and get a piece of plate glass up to say 12 in. by 9 in., have the edges rolled and polished to avoid risk of cut fingers, and you will have an excellent surface for a very modest sum. Make a little tray to hold it and rest it on a felt backing. A cover to protect against knocks completes the job.

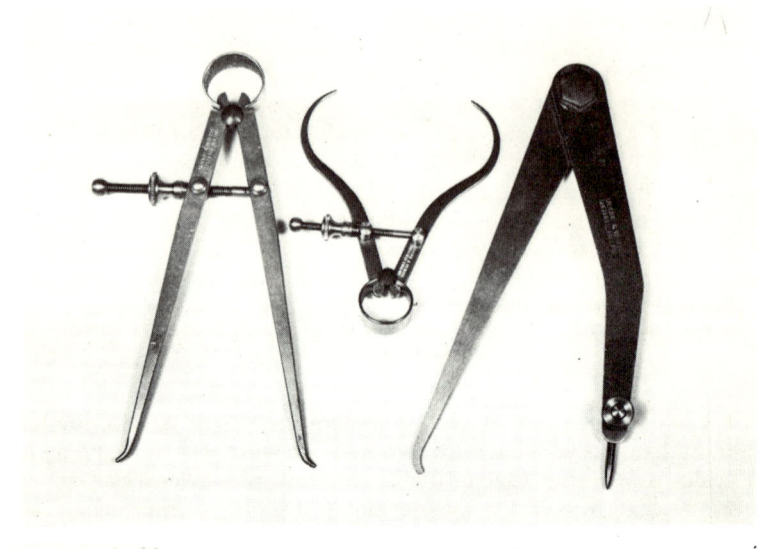

Fig. 5.3 Inside and outside callipers with Jenny calipers, so useful in centre-locating. All from Moore and Wright.

When funds permit there remain the two most expensive pieces of equipment, namely micrometer and dial test indicator. Some fairly modestly priced Japanese 'mikes' are now selling very well. Used with care one will last a lifetime—and as explained above may be acquired luckily through your club. But do try and get one—the mere feel of it in the hand encourages better work! Nothing larger than a one inch will be needed for the Unimat. There is a selection of d.t.i.s on the market. Since they are required for use close up on the lathe a small one is essential. 'Verdict' is an excellent make with its fittings and comes in a neat little case. Perhaps a good idea for a present!

One or two useful items can be made up as practice work. For a start make up a rule stand (this is where the 6in. rule comes in). It consists of a small flat block of mild steel of a size to suit available material (say 1 in. × 1 in. × 1 in. cube). On one side a small hole is drilled and tapped 6BA or similar and a short length of flat spring (ex alarm clock or the like). This with a small limit stop will secure the rule upright to measure from the surface of the surface plate.

Next simple task is a scribing block. Again a suitable block of metal with a dead flat bottom (ours was made up from an old Victorian scales weight). This is drilled to take a short upright which in turn supports a further rod hinged and fixed with a turnscrew.

MEASURING EQUIPMENT

Finally a scribing arm is fixed to this. You can buy it much more elaborately finished readymade if you prefer; or, horrors! make it up ultra simply in wood with a pencil scriber.

With measuring tools, surface plate and work to mark out only a way of seeing the marks must be arranged. This can take the form of engineer's blue (though this rubs off easily) or a chalk paste mixture, simply made which dries quickly and when scribed with a sharp point leaves a clear shiny line. For iron a copper sulphate solution (a few pennyworth of crystals from the chemist) can be rubbed on with a rag and again leaves a bright line when scribed.

Scriber can be easily made up from a short length of round bar drilled up to take an old compass point or even the sharp prong provided free with the asbestos wool used to plug holes in the wall before screwing into them.

Before starting work on marking out file off or rub down any obvious irregularities on the work piece and take one or two rough measurements to be quite sure that there is metal enough and to spare to do the job; also adequate gripping pieces to go in the chuck or accept any other necessary fixing pieces. In general only workpieces of regular shape will go into three or four jaw chucks. Very many items will require to be clamped on the faceplate. Some suitable holding pieces are available from the lathe makers, but in the main, more use will be made of sundry nuts and bolts from the junk

Fig. 5.4 Left to right:—Half-inch micrometer, a real tiny, probably one-inch size a better buy; three fittings for dial test indicator and homemade T-nutted toolpost stand for dial gauge in foreground. A simpler and cheaper device by Verdict can also be obtained (mine was a present!).

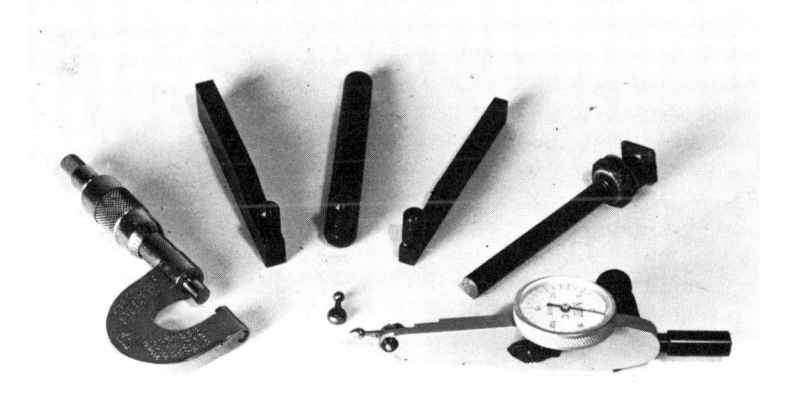

Fig. 5.5 Dial gauge set up on toolpost using home made stand.

box plus a few useful lengths of assorted angle iron which you have cut off in random pieces and drilled to take bolts. Bear in mind that the faceplate is of *light alloy* so avoid over tightening of fixing bolts or it will be distorted.

Be quite certain how you are going to work on any piece. Start with a smooth surface at right angles to any drilling or turning action. This does not necessarily mean facing up a surface. Get it reasonably flat, mount in on the faceplate and then check with the d.t.i. that it is square. Fix d.t.i. in the toolpost by means of one of the attachments provided. Extend plunger just to touch the surface, turn the lathe by hand and note difference indicated. This may be because you have fixed the piece unevenly; if base is firm on surface of faceplate then this is the irregularity to be faced away.

With suitable reference surfaces smooth, marking of centres for drilling and turning can begin. Paint on chalk paste or blue to make lines stand out and scribe up from surface plate. Be sure that you do not remove valuable parts that are essential for holding the work. They may have to come off later, but in due order. It is better to mark more reference lines on the work than needed than not enough!

Next valuable requirement is a rough note pad—rough as you like plus pencil. Note on this reading on dials of handwheels that you will

MEASURING EQUIPMENT   45

be moving. Be quite sure that you know how much a single division movement means. Take a caliper measurement of existing size—in the case of straight turning—note it down. Put under it the finished size required and subtract it from first total. *Half* this amount will have to be turned off (that is the half nearer you and the half at the back of the piece, which added together makes the total!) For example to reduce a round bar from one inch diameter to three-quarters of an inch you take off $\frac{1}{8}$th of an inch all round and it is then the desired $\frac{3}{4}$ in.!

Stop work about two thou. from finished size and take the last two thou. very quietly with a finishing tool to get the best possible surface. In the same with with drilling—if a really accurate fit is indicated the last scrape is better done with a reamer of the appropriate size. Lacking this drill the hole very gently with plenty of 'suds' and if possible use a new drill for the last little turns.

Advice like this may seem very elementary to the more experienced . . . the beginner may find it saves not only a lot of disappointment but also a fair bit of material. Later—sooner if possible!—you may be able to chuckle at this so very obvious advice.

Fig. 5.6   Indicator in toolpost using standard accessory fitting to measure true running of bar in three jaw and supported by fixed steady.

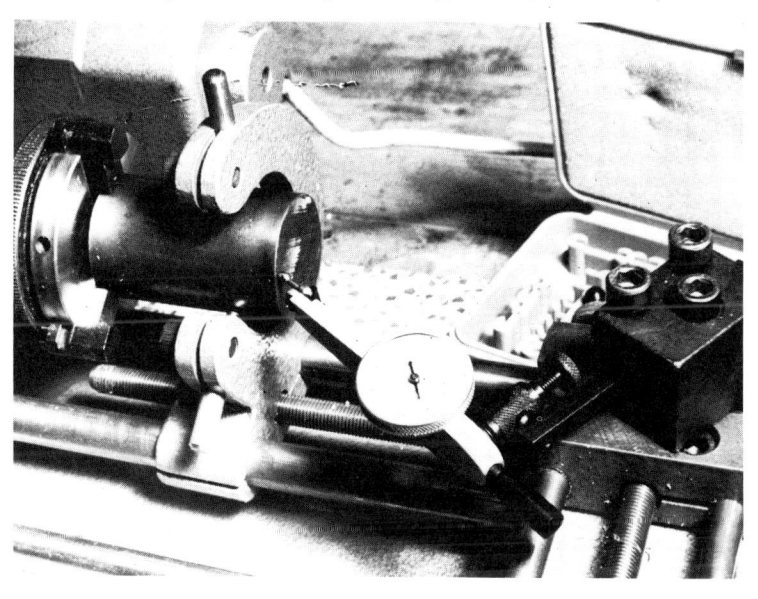

VI. RANGE OF POSSIBILITIES

In describing the Unimat as a universal machine we must accept the limitations along with the manifold advantages. Any machine which is designed to perform a multiplicity of functions must in some respects be at a disadvantage against a number of tools each made for a single purpose. However, over the years users—particularly the non-professional enthusiasts!—have made the centreturning lathe perform miracles of improvisation and do things on it for which the commercial operator would demand a whole factory floor full of apparatus. It just requires a little more time and a lot more ingenuity.

Nearly all functions stem from the original practice of turning between centres. In a small lathe rigidity of work is vital so that

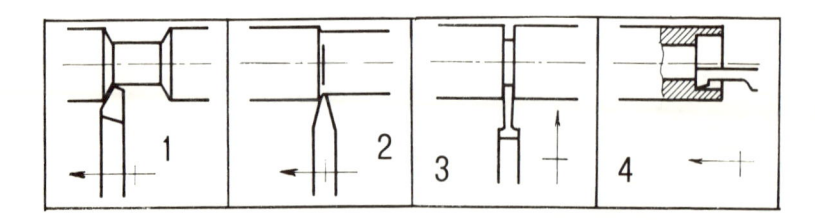

Fig. 6.1
1. Roughing tool
2. Finishing tool.
3. Parting-off tool.
4. Inside turning tool.
5. Right hand side cutting tool.
6. Left hand side cutting tool.
7. Outside threading tool.
8. Inside threading tool.

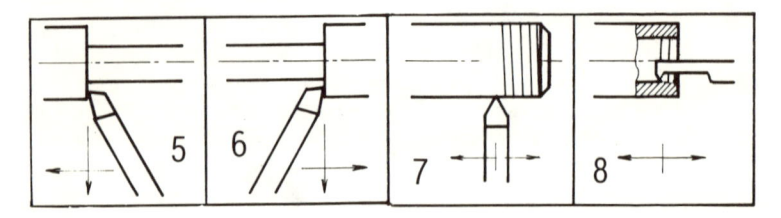

RANGE OF POSSIBILITIES 47

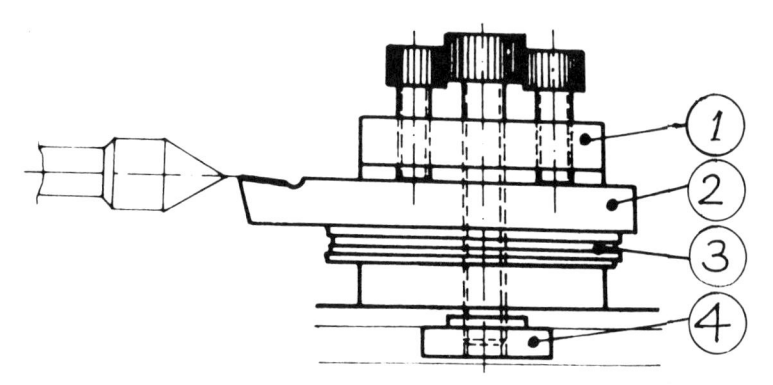

Fig. 6.2 Adjusting tool cutting point exactly to centre height indicated by centre in lathe spindle. 1 Tool holder. 2 Tool. 3 Metal shims (can even be pieces of card!). 4. T-nut.

although improved chuck design has made it possible to support work solely at the headstock end of the lathe where at one time no one would have even considered it, additional support from a live centre located in the tailstock will make machining so much simpler. It is astonishing how much give there is in an apparently firmly fixed and stout metal bar when a cutting force is exerted on it some distance from its point of support. If we consider the fulcrum effect this is not really unexpected.

At the headstock work will normally be placed in the three-jaw chuck, which centres it automatically, or will be driven by a dog attached to the faceplate, when it will be desirable for a centre (as supplied with the lathe) to be inserted in the hollow mandrel to locate it positively. Ends of the work must therefore be centred (or only the one end in the case of chuck held work). This can be done in several ways. Simplest and most common is to use a bell centre punch, which is a bell-shaped guide with a punch running through where the clapper would be fixed. Alternatively a combination square can be used to mark the centre, or the jenny caliper, which is a pair of calipers with legs of uneven length, can be used.

Finally, with short sturdy workpieces the live centre in the tailstock can be run up against work in the chuck. Once punch marked the little centring drill should be chucked in the tailstock and used to make a hole as deep as its countersink. This will house the live centre securely, and a similar hole where needed at the other end will take care of the driving end. A ball-bearing live centre is available as an accessory and is well worth obtaining if any amount of between centre work is envisaged.

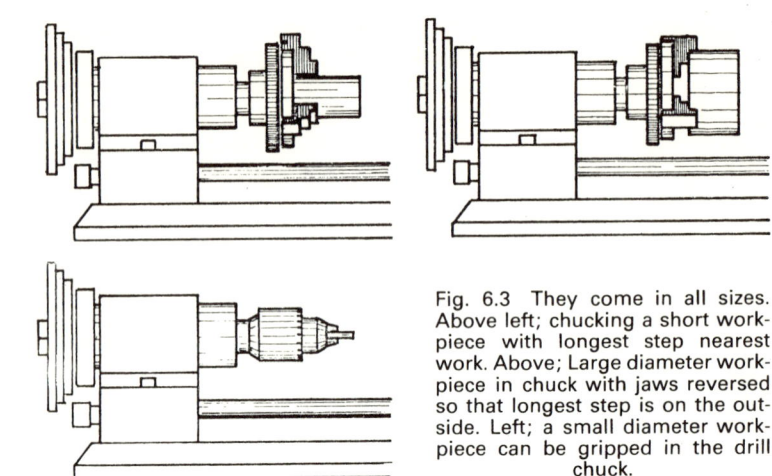

Fig. 6.3 They come in all sizes. Above left; chucking a short workpiece with longest step nearest work. Above; Large diameter workpiece in chuck with jaws reversed so that longest step is on the outside. Left; a small diameter workpiece can be gripped in the drill chuck.

It almost goes without saying that the work ends should be faced off in the lathe. If not possible then at least should be fixed in a vice and filed as flat and as near to right angles as possible. Sometimes none of these excellent schemes are possible, and there is a long whippy rod to centre. This you can *hold steady in your hand* to get the first little centring hole drilled. Wear an old glove if you do it this way, and make sure the toolpost is slipped off the crosslide to avoid catching your hand on it. Do not forget that some degree of hollowness in the mandrel enables part of the work to be slipped up it to secure the remainder more firmly.

Perhaps one of the simplest items that will be required to be worked between centres is a largish bolt, where metal will be removed to leave a larger diameter head. It may finally require to have some threads turned on it. Should there be two such bolts required then they can be made head to head—leaving a little extra for parting off between them. This may be done with the parting tool, or the less skilled may prefer to part off by cutting through with a hacksaw.

At the other end of the skill scale we would place the task of machining say, a four-throw crankshaft, though this is hardly likely to be in the realm of the Unimat except in a very small size!

Far more usually work will tend to be such things as cylinders or crankcases where the amount of overhang from the chuck or faceplate renders them insufficiently rigid without support at each end.

In chucking workpieces remember that the jaws are reversible and that work of a diameter up to 2 in. can be held with the longest step on the outside. Work up to about an inch in diameter can be

RANGE OF POSSIBILITIES 49

chucked with the longest step nearer the centre. Do not tighten too much—it is possible to distort quite solid looking pieces! Very small diameter work up to capacity of the drill chuck (either $\frac{1}{4}$ in. in case of Model SL or $\frac{5}{16}$ in. with Mark 3) can be secured therein. If collets are available these are ideal for very small work, but require the special collet holder. Various sized collets may be obtained but can add to quite a considerable outlay, being stepped in $\frac{1}{64}$ ths through to $\frac{5}{16}$ in.—or in an equivalent metric run.

Either the three-jaw or four-jaw chuck will most likely prove the most used work holder which will support their loads unaided when short and of large diameter cross-section. It is as well to note that the four-jaw chuck when its jaws are extended to their limit is in danger of fouling the lathe shears. This means that the $\frac{3}{4}$ in. thick extension piece must be fitted to raise the headstock. Tool height is then too low and provision must be made to raise toolpost level. Two answers: (1) clamp the toolpost in the machine vice (2) fit the toolpost on the milling table, which in turn is placed on the crosslide. Answer (1) limits toolpost movement and (2) requires the toolpost to be packed up approximately $\frac{1}{8}$th inch and the socket head cap screw for the T-nut changed for a slightly longer one. Note that some degree of rigidity is lost here, so that all non-moving parts should be screwed down as tightly as possible!

In so using the extension piece it should be remembered that the original purpose of this accessory was to accommodate a larger circular saw for woodworking so that use for turning is an extra.

Fig. 6.4 Work attached to faceplate for face or linear turning.

1. Workpiece 3. Faceplate   4. Lathe spindle.
7. Tension screw 8. Headstock

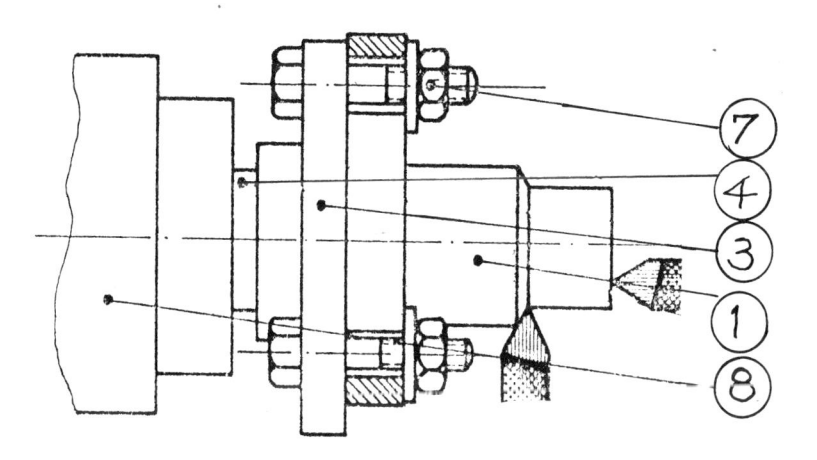

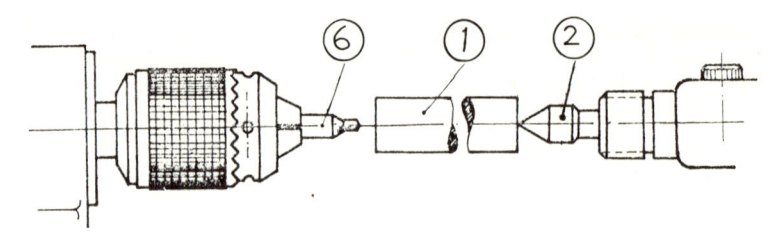

Fig. 6.5 Work being centred with centring drill in drill chuck prior to running between centres. 1. Workpiece. 2. Centre. 3. Centre drill.

(Another use for this extension piece is to enable larger workpieces to be milled or drilled when the lathe is used as a vertical drill).

When only occasional use is to be made of the extension a suitable size of hardwood block drilled and clamped in place will do the trick. No accessory is available to raise the tailstock, so that if this has to be used then an appropriate block must also be made. Purists will be after my blood when I suggest that this too can be fabricated from wood. But do it quietly and you will get away with it. The wood may compress or shrink or not be exact so check centre to centre if accurate work is required (ideally oak or teak is recommended).

The universal nature of the Unimat will soon be tested when tools need grinding. This is dealt with in a later chapter. Grinding is also a function of metal work. A cup wheel is available for this purpose and work can be fixed in the machine vice, with the vertical spindle in place, and work feed across with the handwheel. This method is that adopted in full size practice for truing a surface such as a lathe bed of the conventional shape. Face grinding with the cup wheel in a horizontal position will be the more usual use. Movement is here effected by handwheel of the crosslide. Again be gentle in feeding and always be on guard against some fragmenting of the wheel.

Fig. 6.6 Work running between centres and secured on faceplate. Work being driven by faceplate dog. 1. Workpiece. 2. Centre. 3. Faceplate. 5. Dog. 9. Tailstock.

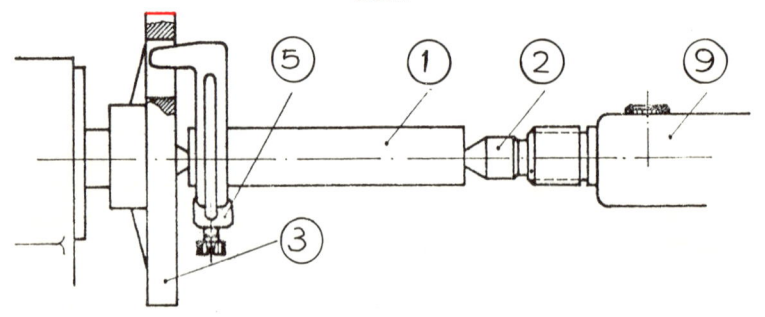

RANGE OF POSSIBILITIES 51

Fig. 6.7 Double header! Reamer in chuck with larger capacity than Unimat's, fitted with arbor which standard chuck will take.

Clean up carefully afterwards—to prevent any grits getting into threads and causing premature lathe wear.

Milling in the lathe is another facet that will occupy a lot of work time and solve many finishing problems. For the most part this will involve the lathe set up with the vertical spindle and the machine vice clamped to the milling table. With larger work pieces it will be found better to fix them directly to the table with the clamps provided. A few additional clamps are always worthwhile and can be bought or made up from scrap pieces of metal, buying only the screws—unless enthusiastic enough to thread up your own bolts of which more anon.

Milling cutters may well be beyond the capacity of the drill chuck and can conveniently be fixed in the three-jaw chuck when over $\frac{1}{4}$ in. diameter ($\frac{5}{16}$ in. with Mk. 3). Alternatively you may shoulder them to fit the smaller chuck since very often the shank is not hardened right to the end. End milling cutters are very like broken drills with their two flat cutters; keyway cutters look more like sharpened spur gears and are side cutters. Side cutters put more strain on the lathe than end mills and should be fed to the work very gently (half a thou. at a time at most).

A fundamental difference between turning as opposed to

grinding or milling is that in the first case the work revolves whilst the tool is fixed and in the latter cases the work is fixed and the tool revolves. In both cases we would say again—and indeed cannot say too often!—that much of the success of the work depends on firm fixing. This will sometimes require considerable ingenuity. Do not be afraid to experiment: the 'book' will not always provide the answer—common sense very frequently will.

Examples of this departure from usual practice could be instanced as using the three-jaw chuck to hold square stock—with two jaws pushing down to hold the workpiece against the third jaw. Then remember that to hold, say, a coupling to bore out a big end one jaw of the four jaw can be reversed; or to hold a rectangular piece two jaws can be reversed. Indeed there are endless possibilities.

Problems will start when the motor is turned on. Work may be very firmly fixed but placed eccentrically. To balance the work so that the lathe is not under strain and sounding as though about to blow up some additional items must be added. This is where the faceplate comes into its own, for, unlike the chucks there is space left to add balance weights. Fat brass nuts, odd pieces of metal, or even slices of mild steel bar drilled and sliced off like Pontefract

Fig. 6.8 Once again reamer too large in diameter for drill chuck so 3-jaw chuck used instead.

RANGE OF POSSIBILITIES

cakes in various diameters should be kept handy for such balancing operations. There is no need to achieve exact balance—just enough to get the motor turning the work over smoothly.

Angleplates may often be required to secure work to the faceplate particularly if machining of castings is invisaged. Castings bought commercially will usually have some chucking piece thereon if intended to be turned. This is cut off afterwards. In any event consider very carefully the series of processes to be done on any such castings so that there is always a way of gripping the work.

Other less arduous but rewarding work will come when using the lathe to polish work. Here a conical spindle is attached to the mandrel by the usual type of flange. With this all kinds of polishing equipment such as felt plates, fabric covered plates, brushes—in fact anything with soft cores and hole diameters ranging from $\frac{1}{4}$ in. through to $\frac{5}{8}$ in. can be used.

Woodworking accessories and methods are dealt with later. Other exciting uses, of course, include a great deal of lapidary work in jewellery making; professional use by dentists mainly on dental plates (not in the mouth, please!) using a flexible drive; opticians, again principally for polishing though also valuable for small fine work on spectacle frames.

VII. GETTING TO WORK

The nominal horse power of the current Unimat is $\frac{1}{10}$th h.p., and its speed 4,000 r.p.m. Under load and with the friction of one, two, or three rubber belts this will drop to 3,450 r.p.m., or even lower and speeds given take this into account. The higher speeds offered for the watchmaker's spindle is because it has a smaller step pulley and tends to be used for lighter work of a high precision nature.

Another factor to reckon upon is a slight variation in the current available from the local power station. At certain times of day, or days of the week, there may be a heavy demand from both industrial and domestic users and hence slightly less power. If it is possible to borrow a tachometer (such as used to test speeds of model aircraft engines) this can be used to check your own motor and see for yourself the effect of different pulley combinations. All motors are not *exactly* the same; they are tested to be within a certain range so hope that yours is among the best of the bunch!

Early models of the Unimat had less powerful motors. My earlier model now used almost entirely for woodwork is rated at only 70 watts (latest is 90 watts) and my original model back in 1956 was rated at only 40 watts. But strangely enough this has never made any noticeable difference to work that I have done! It is just that one gets the feel of a motor, when it begins to labour almost automatically there is the tendency to ease off. If the brutal worker carries on the motor has its own failsafe measure of belt-slip or stall which will follow if too much is asked of it, or, for example a tool digs in or a drill snaps off in the work.

Everyone knows it is hard work to sharpen a pencil with a blunt knife and requires much more effort. The same with a lathe—the smaller it is the more important that settings should be just right, and that tools should be as sharp as possible. This theme will be harped upon throughout the book—it is so very important—and equipment described for skilful sharpening that puts even the beginner on a level with the old hand.

As will be seen the standard motor bracket provides for three

GETTING TO WORK 55

stepped pulleys, a large one on the machine and two smaller ones of similar size, one on the driving motor and one running free. This inter gear or idler serves an essential purpose in enabling progressive step ups of speed. So, we find the idler is brought into use for six of the eleven basic speeds. It will be noted a speed of 2600 r.p.m. can be obtained in two ways, either with the idler or as a direct drive motor to machine. We recommend the latter since it avoids an additional potential power loss from friction.

The range of possible speed combinations available by variations in belt arrangements may prove confusing at first to the newcomer. The commonsense approach that hard metals will demand slow speeds and softer metals benefit from faster speeds is a useful generalisation from which to start.

Next thought is how to relate lathe speeds with the actual workpiece. Quite clearly a thin rod in the chuck will have a much shorter distance to travel in one revolution than a massive two inch diameter bar. For this reason then it is customary to define cutting speeds in terms of feet per minute measured against the circumference of whatever is being worked. Textbooks will give recommended speeds in feet per minute for the various materials in use. These can be used as *guidelines only.* This is important. The Unimat technical booklet that comes with each lathe is commendably vague in this respect! We will try to provide some more positive informa-

Fig. 7.1 Motor bracket with additional idler pulley in use for slowest speed. Note: from motor it is smaller pulley to larger; smaller to larger and finally again smaller to larger on headstock pulley.

Fig. 7.2 Motor pulley bracket with additional idler pulley to enable lower spindle speeds to be achieved. For normal eleven speed range standard pulley bracket is used; or speeds may be varied on this bracket by ignoring extra idler.

tion but please, please, do not regard it as the inviolate rule! It is merely a starting off point.

The nature of the tool steel will also play an important part in the working speed. High speed steel tools can be expected to operate efficiently at speeds as much as double those that can be fixed for common or garden carbon steel. Finally, there is the factor of sharpness. The beginner may fail to keep tools very sharp, may not operate them at the ideal angle to the work . . . It all seems very difficult but do not despair yet.

The Unimat is a most forgiving and helpful creature. If too much is asked of it then it is likely that there will be belt slip. Similarly if speed is about right but too much or too little is being skimmed off then there will be a horrible screaming sound from the tool, and a degree of chatter warning the user not to proceed. By mischance too vigourous a cut may have been taken and the tool dug in. The lathe will stall—so switch off the motor, back out the tool and be less ambitious with the next cut.

Which brings up another point where useful guidance tables can be consulted which suggest suitable depths of cut from various operations. Most of these figures are for use on heavier machinery—remember your little fellow is one-tenth horsepower and be kind. Tables are given intended again as guides not guaranteed routes and take into account the workstrength of the lathe.

GETTING TO WORK

Maker's booklet speeds err on the low side and can usually be speeded up a little. If we come back to the cutting speed per minute and work from that we can arrive at a very fair estimate. Since such speeds are worked out from the formula:

To convert cutting speeds in *feet* per minute to revolutions per minute, multiply speed by 12 and divide by circumference of work in *inches*.

A rather simpler rule of thumb formula to give a quick answer is:

$$\frac{\text{Cutting Speed in feet per minute}}{\text{Quarter of work diameter in inches}} = \text{r.p.m.}$$

CUTTING SPEEDS FOR WORK FROM $\frac{1}{16}$ in/2 in DIAMETER

Diameter of work	Feet per minute								
	15	20	25	30	40	50	60	70	80
$\frac{1}{16}$ in.	917	1223	1528	1834	2445	3057	3668	4280	4891
$\frac{1}{8}$	459	611	764	917	1070	1528	1834	2139	2445
$\frac{3}{16}$	306	408	509	611	815	1019	1222	1426	1630
$\frac{1}{4}$	229	306	382	458	611	764	917	1070	1222
$\frac{3}{8}$	153	204	255	306	408	509	611	713	815
$\frac{1}{2}$	115	153	191	229	306	328	459	535	611
1	57	76	95	115	153	191	229	267	306
$1\frac{1}{2}$	38	51	64	76	102	127	153	178	204
2	29	32	48	57	76	95	115	134	153

*TYPICAL CUTTING SPEEDS FOR HIGH SPEED TOOLS

Speeds in f.p.m.	Soft Steel	Med Steel	Hard Steel	Med Cast Iron	Hard Cast Iron	Bronze	Brass
Rough turning	100	80	60	50	40	60	200
Finish turning	120	90	70	60	50	80	250
Screw cutting	40	35	30	20	15	40	80
Parting off	70	60	50	35	20	50	100
Milling	80	60	45	45	35	80	120
Drilling	100	80	60	80	50	75	200
Boring	80	60	50	50	30	65	150

*Source: Mechanical World Year Book 1975/76
(Aluminium and alloys require between 50% and 100% more speed than Brass)

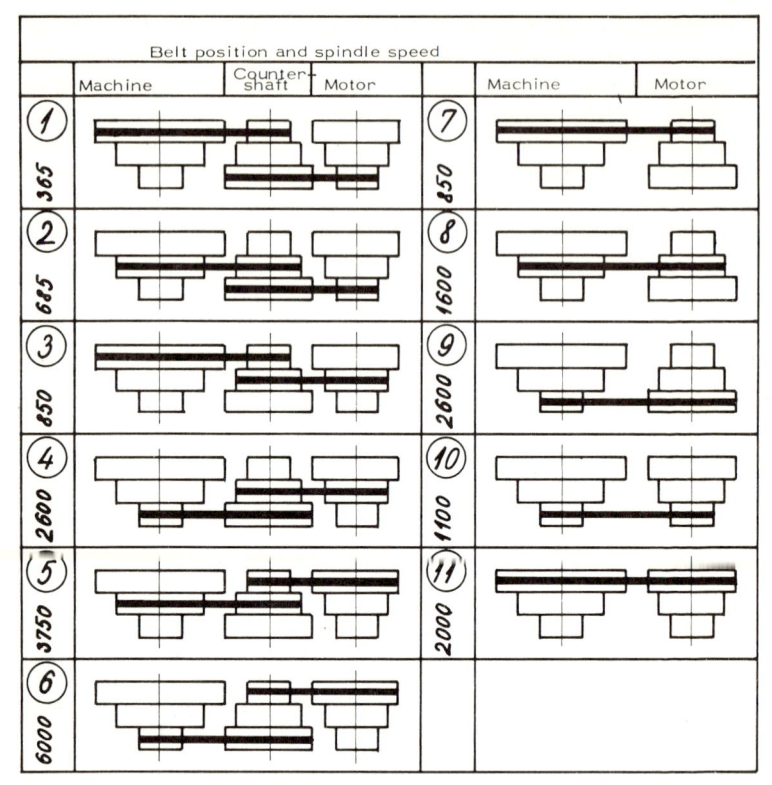

Fig. 7.4 The eleven standard speed arrangements, set out in belt to belt form. Note that positions 3 and 7 give identical speeds as do 4 and 9, using either two belts or a single belt—the latter to be preferred as both simpler and involving less friction loss of power.

Fig. 7.5 Set up for two lower speeds involving the accessory motor pulley and a second idler wheel or countershaft.

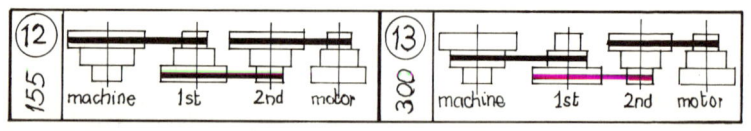

Whether turning or drilling heat is generated and usually calls for some kind of lubrication. When turning between centres the live centre is particularly prone to overheating—listen for a squeak—and here an ordinary lubricating oil, machine oil or similar, is needed. But in the main it is the work itself where the cutting tool or drill is acting on the workpiece that lubrication is necessary. A double benefit is that tools kept cool will last longer and many materials machine better when lubricated.

GETTING TO WORK 59

Traditionally the old turner used 'suds' or nothing more than a mixture of soap and water and fairly slopped it about. This in itself would discourage its use in limited domestic surroundings: it also tends to rust parts of the lathe! A more practical lubricant is soluble cutting oil which is a brownish liquid that turns white when diluted with water in proportion of about one part oil to six of water. This in spite of the added water will not rust or corrode the lathe as the oil sticks to the metal and the water evaporates. At one time it was hard to obtain in small quantities but nowadays many firms stock in 16 oz. or similar size containers.

There may be objection to the smell of the oil if working in the house, but this should not really be necessary if one or other of the proprietary brands such as Shell Dromus (which contains deodorants and antiseptics) is used.

Aluminium and its alloys work best with paraffin oil as a lubricant. Here again, a proprietary such as Cutmax Straight Cutting Oil is a good investment and is used diluted with paraffin. If no lubricant is used then the metal grains cut away tend to build up on the tool and spoil the quality of the finish.

Fig. 7.6 Speed range of Unimat Mark 3 is shown on a plate attached to the headstock as reproduced here. By virtue of the two speed motor only four belt changes are needed to cover the eight available speeds. Belt from motor to idler pulley remains unchanged.

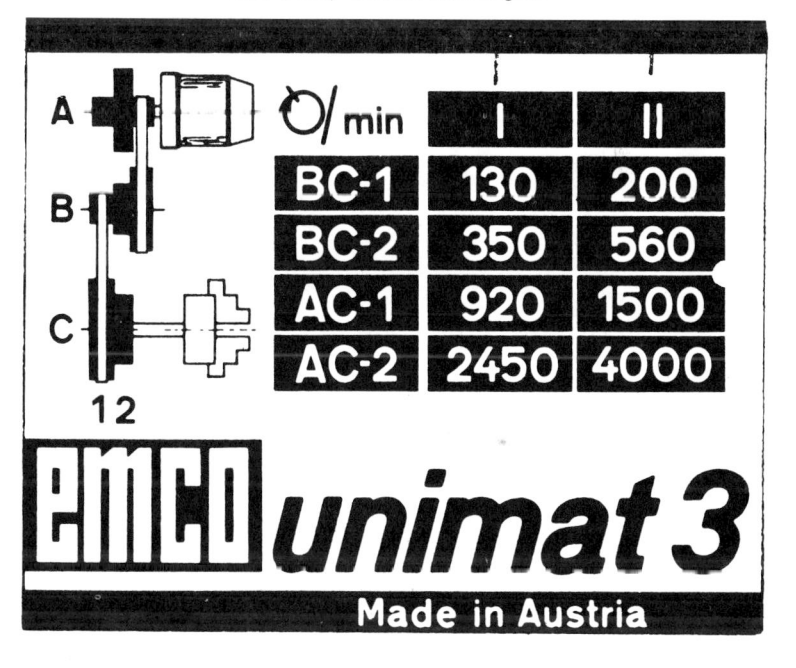

O/min	I	II
BC-1	130	200
BC-2	350	560
AC-1	920	1500
AC-2	2450	4000

emco unimat 3

Made in Austria

Brass and cast iron work better without lubricant, though tallow is sometimes recommended for heavy work on cast iron. For wrought iron and steel lard oil is useful, and molybdenum disulphide as used in motor car lubrication can be added.

With all these operations be moderate—do not slosh on oil generously, but brush it on with a $\frac{1}{2}$ in. paintbrush from a jar kept handy on the bench.

After which it is only fair to add that some turners use little if any lubricant since they claim its presence obstructs their view of progress with the work. This may be fine for the very expert with the sharpest of tools but for less gifted mortals I feel lubrication is the safe answer.

If you are a really wet operator try and obtain an old enamel tray large enough to hold the lathe on its base. This will hold the surplus fluid and avoid any need for large mopping up sessions. An old butcher's tray is ideal—cultivate your butcher.

Fig. 7.3 Self act has been installed. This makes use of larger fixing nut on headstock to secure pulley. It has a further pulley groove thereon for self-act belt. By twisting belt into figure 8 loop reverse action is obtained.

VIII. USING THE TOOLS TO ADVANTAGE

If your lathe is to give good service it is vital that a proper understanding of the tools is gained as soon as ever possible. A set of assorted tools can be purchased at the same time as the machine and most owners will have so indulged themselves. Six tools come in a neat little box and comprise a round-nosed tool suitable for brass turning, a left hand and a right hand tool, a boring tool, a roughing tool and a parting off tool. Note that the *left* hand tool faces to the *right*, that is to say it prepares the left hand of the work; the *right* hand tool works to the *left* to finish the right hand work. These tools come machine ground and sharp but it pays to give them a little touch on the oilstone to smooth out any machine-left roughness.

These tools fit conveniently into the toolholder and are secured by the two screws provided. It is important that the cutting tip is exactly at centre height. This can be judged by advancing the tool to touch one of the dead centres placed in the mandrel. If it is too low then it must be packed up to the centre height with small metal shims, or, failing that with slips cut from a visiting card. With $\frac{1}{4}$ in. square tools it will never be too high. The reason for centre height is so that the tool angle is as designed. If it is too high then clearance for the tip is reduced, if too low clearance is increased and there is greater risk of breaking the tool point. This requirement applies to turning in general but is essential for a tool cutting on the periphery of the work; less important for facing. The toolholder will normally be adjusted so that the tool is cutting at right angles to the workpiece.

It is important that work and tool are firmly secured. Sharp tools, firm fixing and good light are the three qualities to be sought.

The cross-slide on which toolholder rests can be moved to the left—towards the headstock or to the right towards the tailstock by turning the longitudinal handwheel. This is calibrated and each small division moves the lead screw .002 inch. The cross slide has a similar handwheel to move the slide at right angles to the bed back and forth and this is similarly calibrated. Move the tool gently up to

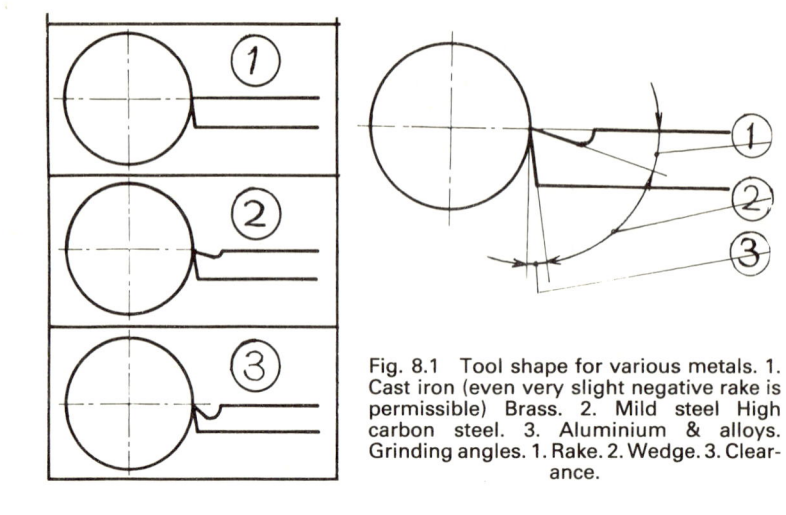

Fig. 8.1 Tool shape for various metals. 1. Cast iron (even very slight negative rake is permissible) Brass. 2. Mild steel High carbon steel. 3. Aluminium & alloys. Grinding angles. 1. Rake. 2. Wedge. 3. Clearance.

the work and take a very gentle cut. It is a good idea to start with brass or alloy for a beginning as the cutting is easier whilst a beginner is getting the feel of the equipment.

We would now digress to consider the precise shape of the tools and the angles at which they should be ground.

Fig. 8.2 Quarter inch square tools available from the distributors. Six come in a neat little divisioned box. They comprise left to right parting-off, inside turning, roughing, round nosed finishing, right hand turning, left hand turning, outside threading inside threading tools. Above is shown a slip of fine India oilstone to keep handy for touching up tools.

USING THE TOOLS TO ADVANTAGE 63

Fig. 8.3 Some special tools. Left is a Stellite tipped tool of ¼ in square—most useful for all purposes and hardly ever needs sharpening. Then three larger section tungsten carbide tipped tools that have been ground down to fit in tool post at centre height. They too keep their sharpness for a very long time.

The action of the cutting tool is that of a wedge which slices off a chip from the workpiece by shearing into it. The angle between the top of the tool and the clearance in front of it is called the wedge angle. The angle at which the top of the tool is set is the top rake angle. There may also be a need for a side rake angle. That in front we have already met as the clearance angle.

According to the nature of the material being worked on these angles will differ. This does not mean that tools will have to be re-set every time they are used but it does indicate that a well equipped operator will have several sets of tools sharpened to suit metals in use. Normally one particular metal is the 'popular' choice and for only occasional use quite satisfactory work can often be done using a tool specifically ground for use on another material.

Tools will—if bought from the manufacturers of the Unimat—be high speed steel (or HSS) which means they can be allowed to get quite hot in use without losing their edge and not that the lathe must be run at high speeds to suit them. A cheaper tool would be carbon steel but this must be used at lower speeds and will require more frequent sharpening. The ideal tool is made from tungsten carbide, usually in the form of a tip only which is brazed on to the shank. This presents special problems in sharpening but happily is of such hard material that it seldom if ever needs a re-grind.

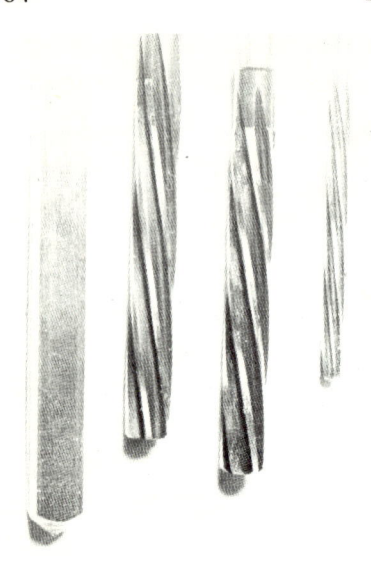

Fig. 8.4 Three parallel reamers with round shanks. Get them in preference to square or Morse taper shanks. On left diamond pointed truing tool for the grinding wheels; also turns plastics and fine turns most metals. Fits tool post though very long.

It has snags of course! First, it is more expensive than other tools; it is also difficult if not nearly impossible to obtain as a $\frac{1}{4}$ inch square tool. It can be easily bought as $\frac{3}{8}$ in. square, rather less easily as $\frac{5}{16}$ in. square. It will then not fit into the Unimat toolholder and be at centre height. It is worth the trouble of grinding down as shown to fit. If any degree of iron castings are to be machined its use is virtually essential.

The problem of castings is first that they have a kind of hard skin which must be removed, and even then the surface is not of a uniform hardness but has specially tough areas. These tend to break off the tip of a HSS tool, which if used should therefore be sharpened with a round nose like a brass turning tool.

Tools can be sharpened on the lathe using the grinding wheels available. Since they have a small diameter they will grind a hollow in the tool if used on the outer edge of the stone. It is better to use the side of the stone which gives a flat edge. If using a grindstone in the lathe remember that the grits can get into moving parts of the machine. It is therefore desirable that the motor is swung at right angles, the lathe shears covered with a cloth or paper protection. The grindstone guard should also be fitted and the careful user will make certain that he is at least out of the line of fire should a stone shatter or better still wear protective goggles.

A very skilful operator can hold a tool in the hand and judge very well the angle at which it should be held—but will in all probability prefer to use some form of jig which achieves that end much more

USING THE TOOLS TO ADVANTAGE 65

scientifically. A very simple jig of this kind can be built up from scrap material and bolted to the wooden stand on which the lathe is fixed during the sharpening process.

The table below is to some extent a summary of expert views on the right angles and like any consensus allows a degree of flexibility. If followed within the limits given results should be satisfactory.

Material	Top rake	Grinding Angle		Clearance
		Side Rake	Wedge	
Cast Iron	0°—2°	12°	78/84°	6°—8°
Brass	0°—4°	14°	84°	4°—6°
Mild Steel	7°—20°	18°	62/75°	6°—8°
Gunmetal	4°	14°	84°	3°
High Carbon Steel	5°—14°	10°	68/81°	4°—8°
Aluminium	40°		40°	10°

There remains the parting tool to consider. This should have a top rake of 10°, front clearance of 5° and side clearance of 2° on both

Fig. 8.5 Sharpening a tool on grinding wheel to which guard has been fitted. Approximate angle can be obtained via taper calibrated dial on headstock base.

sides. But with brass no top rake is recommended. Some users like the actual cutting edge to be angled slightly.

Special considerations must be regarded when sharpening the boring tool. There must be adequate top rake (10° at least) plus some side rake, and good front clearance. Since it will be working in a hole and therefore be slenderer than a standard tool there should be plenty of relief to reduce undue pressure on the actual tool tip.

With a small lathe all turning must be approached with delicacy. No great slices are going to be sheared off—a cut of about $\frac{1}{2}$ mm at a time, say a little more than $1/64$th on an inch should be aimed at irrespective of the metal involved. Once again the advice—SEE THAT THE TOOL IS TIGHTLY SECURED and the workpiece firm. In this way juddering, digging in, motor seizure and chattering will all be very nearly avoided. There is bound to be some chattering, and with practice you will know when this is inevitable and when to stop and rethink tactics.

Practice makes perfect on the lathe as elsewhere. I am not really in favour of 'doodling with metal' but the experts in teaching think differently! The Unimat people publish a most exhaustive illustrated booklet which enables the comparative novice to make some two dozen test pieces which bring into play all the most usual lathe operations. The programme contains the cautionary note that it is intended to be geared to the presence of an instructor. If you are lucky enough to find such a course operating in your part of the

Fig. 8.6 Using a large tungsten-carbide tipped tool on a casting. This speeds up the operation though still a comparatively slow process.

USING THE TOOLS TO ADVANTAGE 67

country our advice is to sign yourself on straightaway and take an enormous leap forward!

Most will have to learn by trial and error. To return to our first cuts we have just begun taking off light shavings from our not too valuable workpiece. Next step is to start taking larger shavings. With a good sharp tool, the right speed and accurate centre height placing real shavings like miniature apple peeling should come off. Still more adventurous cutting and the tool will try too hard to take too much and dig in! This is splendid—now you know how unpleasant it feels and the ugly mark it can make on the work!

In the course of operations you will do this lots of times—usually without too serious a delay in procedure. Sometimes it will stall the motor, whip the piece out of the chuck, take the point off the tool. Nearly always it is the consequence of not paying attention or being too impatient. With the little lathe you can do nearly everything but just not so fast as on big brother.

Parting-off, or cutting off the nearly finished part can be a quite awesome operation. It is—when not carried to the ultimate of actually parting off—sometimes referred to as 'plunge' turning to make grooves and the like, so a little practice on this less skilled operation with workpiece running between centres is recommended to get the feel of the tool. I urge work to be between centres because this tool exerts a great deal of side pressure on the workpiece. As bought it is usually ground quite flat across the end—rather like a chisel. If intended mainly for parting off it is a good idea to re-grind (or ask a skilled friend to do it for you) the end so that instead of being flat it is angled. This helps relieve the side thrust and produces a more gradual parting off, though it may leave a tiny little pip in the centre of the work if not precisely at centre height.

To help relieve the sidethrust some form of support against which the work can be pressed is helpful. With the Unimat cross slide having only the one T-slot this requires a bit of wangling but a support gadget can be rigged up 'Heath Robinson' fashion. Ideally, the tool should be fed from the back so that it tends to force work down by its action instead of the other way up, but alas is not practical here. Some people never do get the knack of parting off and for them we can only introduce the 'dirty tricks' department! A short length of hacksaw blade, preferably in a file handle, held against the toolpost and quite vertical so that it cuts square is pressed against the revolving work and will speedily cut through and provide a parting off. It has the virtue too that it uses less material should that be a matter of importance. It is a confession of failure, of course, but it produces a result.

We should add here another of the 'dirty tricks' repertoire: namely the use of a strip of well-worn emery paper to provide that last finishing stroke to a piece of turning that is properly done with a finishing tool. It is never quite so good but a very close imitation.

Taper turning will be often in demand. With the SL Unimat the head can be loosened and the required amount of taper deflection made by virtue of the calibrated register on the headstock base, tightening up the locking screws. (Early models had no such calibration and with a different type of locating plug could not be so firmly adjusted). On the latest Mk. 3 the headstock is fixed and a limited degree of taper turning can be done with the help of the special graduated top slide. Once the amount of taper has been decided upon the actual turning proceeds exactly as if working on the straight. It is important that the cutting edge of the tool is at centre height or there is the risk of producing an asymetrical cone!

The shape variety of the standard tools has been fined down to a mere half dozen shapes. In any old book on turning simply dozens of shapes with intricate bends and twists will be illustrated. Most of them can be happily discarded. Apart from the straight shapes offered, and here the variations will rest mainly in the angles at which they are ground to suit particular work, departures from the normal are for inside turning where the tool edge comes off at right angles as would be expected and of course a similarly shaped tool for thread cutting in the inside position.

A series of letters to cutting tool manufacturers produced a wealth of information. From this we have been able to try out a number of special materials. For the most part the advantage to the

Fig. 8.7 Useful twist drill gauge for checking lip angle and length of cutting lips. Make from mild steel or alloy, scribing degrees. Twist drill nomenclature. 1. shank. 2. twist drill. 3. flute. 4. margin. 5. cutting lips. Angles marked should be the same—checked with the gauge. Lip clearance angle is marked for fine between 6–9°. Lengths equal.

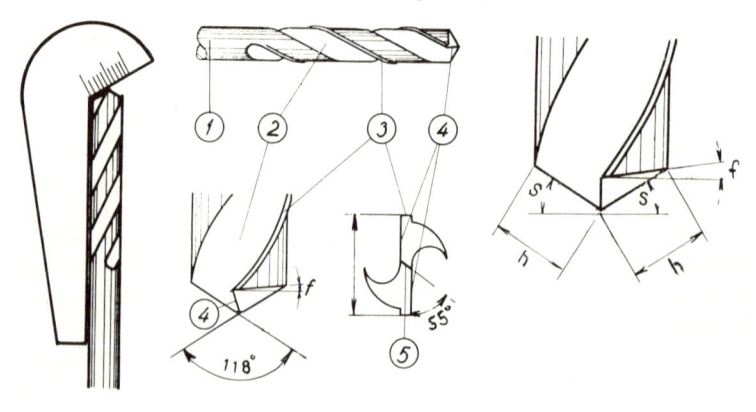

USING THE TOOLS TO ADVANTAGE

Fig. 8.8 Smart new style three-point fixed steady for Mark 3. Inside turning tool in use.

amateur is limited since they tend to be higher in price than the usual run of equipment. We have, however, found the smaller *round* section tools of special interest and value. *Model Engineer* Editor Martin Evans has devised a very neat little holder for such a tool, sharpened as through for a boring tool, and sliding in an inclined hole drilled through a square section holder, and locked in place with a set screw. The opposite end of the tool also has a longer screw, which can be screwed in further as the tool wears and is resharpened to use up the tool blank, rather like pushing up the lead in a propelling pencil! This has great possibilities.

Virtually all turning work boils down to applying a tool to a revolving workpiece. The skill comes in so arranging that the revolving workpiece finishes up in the desired shape. With experience of plain turning it will not be long before the need arises for eccentric turning. Common uses will be for such things as the journals of crankshafts. This can be done by mounting the shafts between their respective centres and treating each operation as a simple turning job.

With, say, a job like the crankshaft for an i.c. engine where we wish to convert the up and down motion of the piston into the circular motion of a propeller it is a matter of very carefully marking out the two sets of centres on accurately faced ends of the workpiece, centre drilling exactly and cutting away. Even here there may be minor problems with the two sets of centres, perhaps the most

Fig. 8.9 Flycutter held in collet holder and mounted in vertical drill to cut flat on bar stock held in machine vice set up with Mark 3 model.

difficult being the need for some considerable tool overhang, to get sufficient clearance. Another factor where a lot of metal has to be removed is weakening one part of the workpiece too much too soon with the possibility of fracture or bending against the tool. Once more the injunction sharp tools, firmly secured and the appropriate speed for the material must be remembered.

Getting a little more elaborate and turning a single throw crankshaft from bar material it may be necessary to have three sets of centres marked out, namely to first turn the edges of the webs, before getting on with the major metal removal. Here it will usually be advisable to cut away with the hacksaw any large areas of metal (this is specially so with a small light lathe such as ours). We soon come to a situation where we are turning the two ends and have cut a gap in the middle between the webs. It is desirable that this gap be filled with a temporary gap piece, held in place with a throw piece. If the gap is small a couple of washers bolted up would suffice. The tool turning the crankpin should not be too wide or be used for taking too ambitious a cut—little and often is better.

Nowadays it is often possible, and certainly preferable to obtain a casting where much of the surplus metal has already been taken care

USING THE TOOLS TO ADVANTAGE 71

of! Original designs can frequently be created round some existing commercial casting, or failing that patterns made up in wood if a friendly local foundry is willing to slip in a casting or two between runs. The enthusiast will soon find out about his local suppliers, particularly if he joins a local model engineering club, or attends sessions at a handicraft centre run by the local authority.

In some instances, even with simple crankshafts, it may be wise to bolt on false endpieces for finish turning of the crankpin. From single throw to four or five throw is an ambition rather beyond the little Unimat. In the small sizes envisaged, if such is really required, it would be possible to build up such work rather than turn from the solid, and still be strong enough to operate satisfactorily.

The simpler operations on the lathe involve a fixed tool which cuts against a revolving workpiece. When we do drilling in the lathe the same state exists—the drill goes in the tailstock and is fed to a revolving workpiece. If the vertical column is erected then, of course the more usual function of a drill to go round and round is observed. This equally applies to milling, where the work is secured firmly and a milling cutter revolves. The important word is *firmly*. To obtain good milling results from a lathe depends on its size and the degree of rigidity obtainable.

It is perhaps significant that only with the introduction of the Unimat Mark 3 has a flycutter been listed as an accessory. However, owners of an SL need not feel inhibited from flycutting if they will follow the improvement recommended by Mr. Tingey of fitting a slightly thicker silver steel bar to their cross slide in place of the existing one on the right handside. The standard bar can easily be drifted out and the slightly enlarged holes reamed out in situ with a reamer of suitable size. Improvement is quite amazing!

We can define flycutting as an operation using a single point cutter mounted in a bar held in the chuck and rotated against the work suitably secured on cross slide or milling table. This means that work is nibbled away by a series of pecks by the single pointed cutter which obviously requires greater rigidity than a continuing cutting motion. Milling cutters are quite expensive and a series of shaped cutters could present a lot of investment, perhaps for only occasional use.

Happily a flycutter can be made up by grinding a short end of a broken drill. A hole of similar size to the drill stub involved is then drilled in a short length of bar that will conveniently go in either drill chuck or three-jaw—drilling angle being about 45°. Another small hole is drilled at right angles and threaded to take an Allen screw, which secures the tool.

From this simple basic notion a whole team of useful tools can be devised. To double up on speed two tools can be fitted to the bar to provide two cuts per circuit. More usual, however, is its use as part of a boring bar. With the little Unimat this is a very helpful accessory. Here a bar, which can be of as small a diameter as ¼ in. is mounted to run between centres. The cutting tool is fitted either at an angle (say 45°) or can be at right angles and duly secured. It can now be used to bore holes very much larger than can be tackled with the normal boring tool in the toolpost. Even when it could be done as a turning operation it is easier to set up and maintain accuracy with work mounted solidly and not revolving. In many cases it overcomes problems otherwise insurmountable, as for example boring a hole in a workpiece too large or too awkward a shape to be swung in the lathe.

The secret of success is to take very small cuts and so not overtax the lathe, mere scrapings of .002 in. being suggested for mild steel, with a little more, say .005 for alloys and brass. Speeds should be slowest possible and always feed *against* the direction of rotation. If you attempt to go with it on a small lathe the tool will tend to take over and grab at the work. You need only try it once to be convinced! This rule applies to all small lathe milling work, whether with fly cutters or normal multi-toothed milling cutters.

Further variations on the theme include shaping your own cutters to produce desired shapes. Gear cutting is the most usual such need, or V-shaped keyways. T-slots would normally be done in two steps, a first slot to represent the pillar of the T, and then a cutter to make the cross-bar. The use of slitting saws for milling operations should not be overlooked. They can be used singly or with two or more together as gang-millers. We have mentioned elsewhere that surplus blades can often be bought very cheaply, provided you are prepared to make suitable adapters for unusual centre holes.

In the course of time you will probably build up several cutter bars of varying diameter to deal with different sizes of work. The only tedious part of the job is really the need to adjust the cutter frequently via the Allen screw holding it in place. Have enough length to the tool to allow for such extensions—though the shorter the protrusion of the tool the more rigid it is. It must not be so long when used as a purely boring tool that the rear end fouls the work. This is really the case for having several bars made up.

With boring between centres difficulties of mounting work will mainly be concerned with firm fixing. With slotting work it will often be a problem to get work located just where it is wanted, because of centre height. The vertical drill column and milling table answer this question very successfully. The Mark 3 offers an additional fitting to improve even more the useful functions of this vertical column.

USING THE TOOLS TO ADVANTAGE 73

Fig. 8.10 Beginning to turn a crankshaft casting. Work is running between centres, using standard faceplate with dog and ball bearing centre. Tool has not yet been advanced to work. Material between webs has been left in for strength.

Fig. 8.11 Turning a crankshaft on the Mark 3. Note long tool to reach to work. Temporary throw plates have been screwed on to ends of work to enable it to run between centres.

If you have been able to rig up a vertical slide many problems of endmilling will be solved. In the absence of such a useful accessory, remember that small workpieces can be set vertically on the machine vice with the aid of a piece of angle iron; the faceplate can also be secured on the cross slide to provide a platform; the tailstock can be advanced and hold the faceplate as a vertical surface. This last should be tackled with caution as the tailstock is not really robust enough to take a lot of stress, so cuts should be very small, and screws all thoroughly tight.

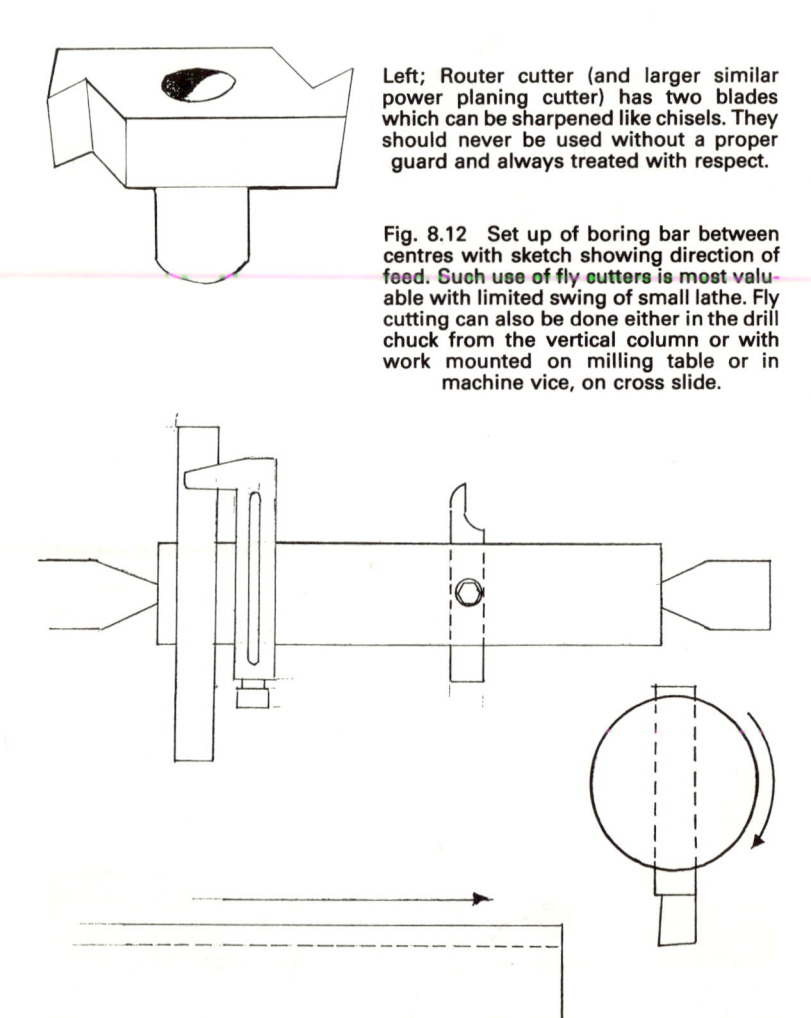

Left; Router cutter (and larger similar power planing cutter) has two blades which can be sharpened like chisels. They should never be used without a proper guard and always treated with respect.

Fig. 8.12 Set up of boring bar between centres with sketch showing direction of feed. Such use of fly cutters is most valuable with limited swing of small lathe. Fly cutting can also be done either in the drill chuck from the vertical column or with work mounted on milling table or in machine vice, on cross slide.

IX. SPECIALIST ACCESSORIES

Sooner or later the question of screw-cutting will be raised. There are three ways in which screw threads can be cut on a lathe; which method is chosen will depend on the size of thread, the general need for a lot of thread cutting, and the ability of the lathe to carry out the work.

Simplest method of all, and very suitable for small sizes of bolts and nuts is to rig up taps and dies in the lathe. In fact a lathe is not essential at all, but its availability makes the task much easier and rather more accurate. The work can be secured in the chuck or collet at the headstock and a die holder fitted to the tailstock which is brought up to the work. The mandrel is then turned *by hand* to feed into the die, loosened off as much as possible. It will begin to bite and must be fed on very carefully, and then turned back the other way to clear metal chips. Continue the process until sufficient thread has been formed. Close up the die and go through again until adequate depth of thread has been cut. Test by running a nut of appropriate size up and down. Lard oil is a good lubricant for steel, or any of the oils normally recommended for turning lubrication.

Bar that is to be thread cut in this way can be turned very slightly undersize of the finished screw, since the action of the die distorts the metal slightly and makes good this undersizing.

Holes can be tapped in exactly the same way using the appropriate taps; again progressing carefully from the smallest size and not attempting too fast a cut; turning back to release chips and giving adequate lubrication.

Small sizes, particularly with threads such as the B.A. (British Association) sizes, are nearly always made on the lathe in this way.

The expert will now figuratively brandish his fine lathe and tell you it is a screw-cutting lathe. Quite right; if the lathe has a leadscrew that can be coupled to the headstock by means of gearing then by making use of the lathe changewheels (a set is 22 strong) virtually any combination of lead can be made up. This alters the rate of progress of the cross-slide along the lathe bed to match exactly the pitch of the screw to be made. In other words if the t.p.i. (threads per inch) needed are say, 24 t.p.i. then the tool on the cross slide will advance exactly one inch while the work is turned 24 times. Setting up such a chain of gears takes an appreciable time and demands a degree of skill and know-how from the operator.

On the Unimat the designers have gone straight to the method employed on production machines by providing a hob or leader accessory which leads a chaser along the work to provide a perfect thread. As with a set of changewheels it requires to be set up on the lathe, but if any great quantity of threaded work is required then it is a desirable accessory. Each size and thread form requires its own hob and chaser so that a quite frightening collection of hobs and chasers (or leaders and followers) could be built up by the deep of pocket. However, since most of the smaller sizes will be cut with taps and dies, up to say a diameter of half an inch, a more modest selection can be made to serve all practical needs.

With the advent of metric sizes and the accelerating obolescence of old friends like B. A., Whitworth, A.N.C. and the like, it is prudent to consider standardising on metric from the start. Since a number of items will almost certainly be made to suit the lathe—accessories and so on—it is useful to realise that all threads on the Unimat are 1 mm pitch from the bolts holding parts together right up to the sindle head sleeve. There are metric pitch hobs and chasers from 0.5 mm up to 1.5 mm in regular steps, and with 0.75 mm and 1.25 mm., but choice of the 1 mm will provide a useful and extensively employable unit. Later if other needs arise then additional sets can be ordered. It is valuable to obtain a metric thread gauge as soon as possible to avoid error. I also have a set of the sizes up to 12 mm which in the usual form do not follow a single pitch but work up in steps according to diameter.

First sight of the accessory and first glance at the fitting instructions are somewhat forbidding. Do not at this stage attempt to install it on the lathe—just look at the parts and decide which piece is what. The cutting arm will undoubtedly have the internal threading toolholder attached to it, which is momentarily baffling. This part is a nearly round rod with a stout U-piece at one end. Release this and the purpose of the cutting arm to hold a standard tool becomes evident!

Next unscrew the backplate from your three-jaw chuck. Once unscrewed it comes away from the part very easily though maybe a little stickily. This is now slipped up the leader so that the threaded flange protrudes slightly, and the three slightly longer screws provided are used to screw the chuck to that end. The complete chuck is now attached to the leader, and the whole unit is located on the pushed out spindle head sleeve, clamping the sleeve firmly with the

SPECIALIST ACCESSORIES 77

Fig. 9.2 Screw cutting hob set up for internal threading. Cutting arm is pressed on the toolpost carriage to hold tool in position as it is cutting at a very slow speed.

Fig. 9.3 Another shot of the same operation which clearly shows the action of the cutting arm in holding tool in engagement.

two clamping screws on the headstock.

Large diameter work it is clear can be secured at the headstock end by the three-jaw chuck in the usual way. Should it be desired to cut external threads between centres without clamping in the chuck the chuck jaws can be removed (unwound), a centre slipped in the hole in the headstock and a lathedog secured to the work and lodged in one of the jaw guide grooves. This is what it says in the instruction booklet! I think it unkind to poor chucks and not guaranteed to ensure their long life and would only follow this suggestion in the case of the direst need. You have been warned.

Next stage is to fit the two supporting brackets on the lathe shears. One at each end—taking care that the headstock end bracket is clear of the knurled ring of the lathe chuck. The tailstock end bracket should be as far to the right as possible. If work is to be supported between centres for turning it should be to the right of the tailstock. If work is perhaps shorter and can be supported with the fixed steady then the whole tailstock can be unscrewed and put on one side. The guide rod is pushed through from the tailstock end—the spring being kept to the right of the tailstock bracket. The cutting arm goes in the middle so that the spindle knob lies just over the cross-slide; the guide arm to which the little follower is attached goes on the left hand end of the rod to the left of the headstock supporting bracket.

By this time it should be fairly clear what it is all about! A tool goes through the hole in the springloaded device on the cutting arm. The follower must be adjusted on the leader so that there is no play—any that there is will be reflected in the poor thread cutting!

A test piece should now be tried by inserting a brass or mild steel rod of say ½ inch diameter in the chuck, supporting between centres or by means of the fixed steady. Direction of feed is towards the headstock. The spindle knob will be resting on the slide rest and this should be adjusted so that it is to the right end of the rest. A suitable tool having been inserted (Note that metric threads require a tip angle of 60°, Whitworth 55°); the handwheel on the cutting arm adjusted according to the diameter of the work and the spindle knob wound down so that pressing on it engages the tool on the workpiece. Adjust cutting speed to suit size and material and we are ready to make a start. Be sure that everything is taut and ready to start.

Lift the cutting arm clear of the work, start the machine, lower the cutting arm until the spindle screw touches the slide rest and is allowed to slide over it as the running machine feeds the tool towards the headstock. It should be quite clear what is happening—the follower which is clamped to the guide rod directs the tool to follow the same pattern that it is following on the hob. The

SPECIALIST ACCESSORIES

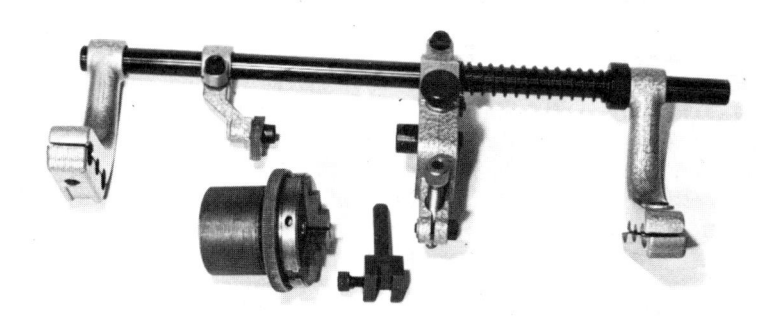

Fig. 9.1 The threading attachment parts. Support brackets are shown at each end attached to the guide rod; cutting arm is in the middle with the spring and thrust ring. To the left foreground is the leader attached to the three jaw chuck and to its right the internal tool holder. When on the lathe the guide arm and chaser seen above the leader goes outside the bracket on the extreme left; similarly the spring and thrust ring goes outside the bracket on the extreme right.

guide rod being free to move in its two supporting bracket holes leads the tool towards the headstock. Then as the first traverse is completed the guide arm is lifted from the work and the spring at the right pulls it back to be ready to start a second traverse. Three such runs at the initial cut and the little handwheel at the top of the cutting arm can be turned to increase the cut by about .005 inch; three more runs and increase again until sufficient depth has been gained. This is checked with the thread gauge or rather more satisfactory a nut of matching size can be run up on it. For long thread cutting runs remember that the slide rest must be wound along as need be to give support to the spindle which acts as diameter adjustment thread. This is careful work and should not be hurried or scamped.

I am not really certain that the collet chuck should be regarded as a luxury accessory! It comes with a separate backplate to which it is screwed and this threads onto the mandrel in the usual way. Unlike most other accessories the holder and locking ring have a metric pitch of 1.5 as against the more general M1 pitch of the lathe and its other tools. The collets themselves are very beautifully made, and it is here that I think the luxury comes in. They are much too good for the normal use likely in a home workshop.

The 'standard' collet is tubular, with a taper on one end, and is slit in three cuts lengthwise. In the centre a hole is drilled to take a given size of drill, workpiece, or what have you. The slits are compressed by screwing down a locking ring on the holder. They have the advantage of being more accurate than a drill chuck though employing a similar principle. The Unimat collets are more delicately made with a double taper and six slits instead of three longitudinally. They are available in size $\frac{1}{32}$ in. (.5 mm) to $\frac{5}{16}$ in. (8 mm) in steps of $\frac{1}{32}$ in. (.5 mm) and *cost several pounds each*! so that a set would cost very nearly half the price of the complete lathe.

Nevertheless they *are* most useful, and I was delighted to re-read an article by my old friend Laurence Sparey where he describes making collets of all sorts of materials even mild steel. You will still have to buy the holder and I would recommend a single collet as a guide.

Another non-essential but very useful accessory is the polishing spindle which comes, again, with its own backplate to screw directly on the mandrel. It is tapered and will thus take an assortment of polishing mops of various centre holes. Alternatively a simpler polishing holder can be devised using the grinding wheel holder. This may not be so versatile as the spindle but can serve very well. In this connection I know one very skilled model engineer who makes beautiful model yacht fittings, has them plated, and polishes them on the Unimat to a quite remarkable degree of finish.

The watchmaker's collet sleeve and the flexible shaft are items which are either essential to one's work or entirely unnecessary so the pros and cons do not really arise. It should be noted that the watchmaker's spindle has its own special pulley which enables it to operate at rather faster speeds than otherwise possible on the standard pulley.

In the woodworking department there are two accessories which will demonstrate their need entirely dependent on your special interests. These are the planing attachment and the routing attachment. For the ship modeller or the architectural modeller they will rank as musts. In general they are excellent for *small* work but require quite a lot of improvisation if large work is to be attempted.

A separate and universally useful accessory is the circular table. This can be secured to the indexing attachment in the same way as the chuck with the countersunk screws provided. It is a substantial table of cast iron with three symetrical T-slots and four concentric grooves scribed on it to help in locating work centrally. There are also three additional threaded holes (M6) to hold clamping plates provided. Its particular function is to act as an extra part for the indexing attachment and will permit of irregularity shaped workpieces being attached for indexing.

Like so many Unimat accessories however, it has other uses and

SPECIALIST ACCESSORIES

will be favoured by all those who find the alloy faceplate too light in structure for their needs. It is worthwhile obtaining an additional backplate for permanent attachment if regular use other than with the dividing head is expected.

So far we have spoken of fairly advanced uses of the indexing devices for gear cutting and the like. It should be remembered that it also has more mundane and useful abilities such as milling flats on nuts (making your own T-nuts from round stock for example!) or making square faces for ends of bolts, hexagon cutting, slotting for keys, grooves and so on.

With all uses remember that firm fixing of parts is essential. Action with a well oiled index gear is very smooth so that some doubt may be experienced as to how many gear slots have been moved forward. However, by glancing down the groove in the clamp slot, that is the slot at the top of the device they can be seen clearly—though it may be necessary to adjust the working light to shine down the slit. Once again, may we remind users of the value of that scrap pad on the bench? Here a five-bar gate type of tick off of spaces moved is invaluable. All the movements should be practised without anything valuable set up to acquire manual dexterity, without risk of spoiling.

Users of older equipment will notice that slots instead of holes are provided for holding down bolts of the indexing head and the detent casting finish is slightly different. It makes no difference to the working.

INDEX GEAR TABLE

Number of teeth	Index Range
24	2, 3, 4, 6, 8, 12, 24
30	2, 3, 5, 6, 10, 15, 30
36	2, 3, 4, 6, 9, 12, 18, 36
40	2, 4, 5, 8, 10, 20, 40
48	2, 3, 4, 6, 8, 12, 16, 24, 48

The need to space out work radially will probably arise. Rough and ready marking out can be done with jenny calipers, dividers and pure measurements but this is imprecise. Whether the need justifies the acquisition of proper dividing or indexing equipment depends on the user. If any amount of gears are likely to be made or there is need for regularly spaced holes in, say, a cylinder head, or there are evenly placed slots to be milled then indexing attachment will be needed.

With a larger lathe operated by a train of gears it is usual to adapt

Fig. 9.4 Self-act or power feed which shows engagement of special feed handwheel with worm; worm at other end of shaft engaged with power feed shaft driven from the headstock end. Disengagement is affected with the ball knobbed lever on left.

them for dividing purposes since a commercial worm driven attachment is expensive. This would normally allow an almost infinite range to divisions to be selected by a variation in the train of gears. The Unimat device employs an indexing head into which a gear (very like those from a lathe gear train mentioned above!) is fitted. This has a stated number of teeth ranging from 30 to 48, the latter being that supplied with the accessory. A sprung detent or plunger locates between the gear teeth holding the wheel in place. It can be lifted and a desired number of teeth allowed to pass. Thus, at its simplest, if we divide a circle into three for say cylinder holding down bolts, then we start at zero, move on sixteen teeth for next hole, another sixteen for the second and we are then back at the start, since 48 teeth divide into three. Any other combination can be made with the useful 48 teethed wheel in the range 2, 3, 4, 6, 8, 12, 16, 24, 48. A 24 teeth wheel would be nearly as embracing. It will be noted however that useful metric divisions allowing 5 or 10 spacing is not possible. For this a separate gear is required with 30 or 40 teeth. Divisions like 7, 11, 13, 17, 19 are not exactly possible with the device and hopefully will be avoided. Such dividing operations have exercised the ingenuity of mathematicians for a long time and provide a teaser for the expert!

SPECIALIST ACCESSORIES 83

Fig. 9.6 The stout clamping plate with four T-slots (three only on SL version) holding work in place with clamping shoes as offered for the Mark 3 Unimat.

Fig. 9.5 Indexing attachment set up on the toolpost carriage of the cross slide attached to the three-jaw chuck. This holds a gear which is being cut with a module fixed in the circular saw arbor with lathe headstock on vertical drilling spindle.

The index gear plate is held in place with a circlip. To remove it easily a pair of snap ring pliers should be used. These fit into the little holes of the ring and constrict it for lifting out. They can be made by softening an old pair of pliers, drilling the top faces and driving in short metal pins to fit the holes. Action in use will be awkward pull at the handles instead of squeezing, but for only occasional use save buying another tool. Watch out the rings are springy and can easily fly off into a dark corner!

This covers a description of the basic accessory. It is versatile in its application. It can be attached to the transverse slide using the two tee-nut bolts supplied. Alternatively it can be set up attached to the normal three-jaw chuck and located vertically on the cross slide so that it can be operated to enable a given number of gear teeth to be cut using a module cutter in the vertical drill column. Be sure that the indexing head is screwed up tight to avoid any possible slipping since work will actually cut during the operation.

Fig. 9.7 Polishing spindle with two mops attached: these can be fitted in a variety of sizes and materials.

X. THE LATHE AS A DRILLING & MILLING MACHINE

When drilling holes in the lathe the workpiece is set to revolve and the drill remains still. This imposes limitations on the shape and size of the object to be drilled. With the vertical column duly erected and the headstock duly transferred to the upright position drilling can take place with the workpiece held still on a horizontal plane.

But first let us consider any special problems of installing this change of user. The column fits into the hole vacated by the headstock and is secured by the same tension screw that retained the latter. At the other end of the column is a clamping support which holds the whole headstock/motor/chuck unit. This is free to slide up and down the column being then fixed at the desired height by a clamping screw. The headstock enjoys the same ability to move round that it did as part of the lathe (i.e. for taper turning) and if upright drilling pure and simple is required then the alignment marks must be adjusted accordingly. Needless to say the drill chuck will be fitted in place and the pinion fitted in place to rack the drill to the work. A return spring for the pinion handle can be obtained if desired. Where a larger drill than the drill chuck can accommodate is to be used remember that the three-jaw chuck can take it in lieu of the drill chuck.

To be quite sure that the drilling apparatus is truly vertical thread the faceplate on to the headstock spindle and lower it to the cross-slide ensuring that it lies absolutely flat thereon. Any small discrepancy will be only too evident.

The faceplate can again be brought into use as a boring table securing it to the slide with the slotted screw provided. For round work the three-jaw chuck can serve a similar purpose. But the milling table will be the more useful accessory here in a majority of cases. Longitudinal and cross movements of the slide enable very accurate location of work for drilling. Do not forget that coolant is just as necessary with the drill as in turning. Suggested drill speeds are tabulated below according to diameter of drill in use:

	Steel		Other metals	
Diameter of drill	rpm	belt no.	rpm	belt no.
1–3 mm	1600	8	3750	5
3–4 mm	1100	10	2000	11
4–5 mm	850	3	1600	8
5–6 mm	365	1	685	2
*6–8 mm	350	–	560	–
Centre-drill	850	3	1600	8

* With Mk 3

Figures by courtesy Emco Basic Course Instruction Manual

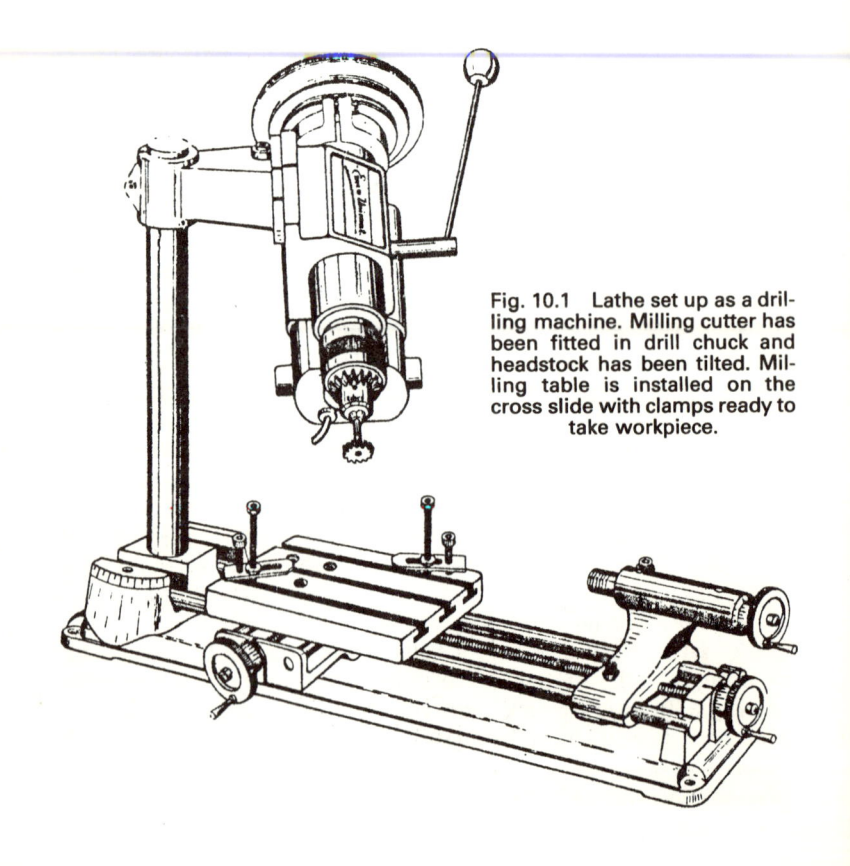

Fig. 10.1 Lathe set up as a drilling machine. Milling cutter has been fitted in drill chuck and headstock has been tilted. Milling table is installed on the cross slide with clamps ready to take workpiece.

THE LATHE AS A DRILLING & MILLING MACHINE 87

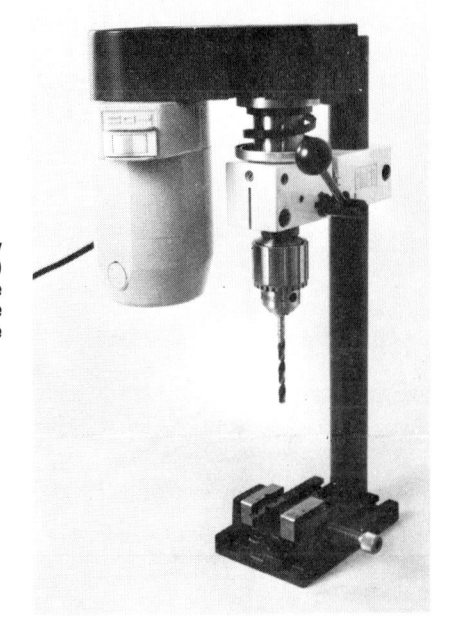

Fig. 10.3 Posed set-up to show fitting of vertical column (Mk. 3) on milling table with machine vice in place for the work. Table would normally in use be attached to cross slide.

Fig. 10.4 Vertical column mounted on the cross slide and carrying cutting module. Work is held between centres with Mark 3 indexing attachment in the lathe. This toothed pinion can later be parted off to provide spur gears or could be used as a locking handle for the lathe or other work.

Fig. 10.2   Earlier type SL lathe headstock on column for vertical drilling. Note smaller size of headstock adaptor fitting. This is of heavier metal than the later and larger design which offers more bearing surface.

Fig. 10.5   The very useful circular saw arbor carrying a metal cutting saw being used to part off work held in the machine vice. No problem with either model.

THE LATHE AS A DRILLING & MILLING MACHINE 89

When using the SL Unimat do not forget the very versatile accessory that raises the height of the headstock by ¾ in.—originally offered to enable a large diameter saw to go on the circular saw —which can be used to extend the reach of the drill in exactly the same way. Be specially careful, however, with this to check that the column remains perpendicular since the added overhang weight tends to pull it out of line and screws must be well tightened.

The full value of the vertical drill will only be realised when all the possibilities of the accessories have been considered. The machine vice is of great value in holding round work where for example it is desired to drill through the perimeter of a ring. It also has a suitable V-groove so that a pipe or bar can be held for perimeter drilling.

Drilling at an angle can also be achieved by swinging the head and locking at the desired angle. This is particularly useful with the later Mark 3 model where the angle is marked, but with reasonable care there is no need to be any less accurate with the original model. Repeated advice: keep a notebook on the bench and write down settings to enable exact repeats!

As there is a limited reach of the drill via the pinion lever it is possible to drill to a required depth without the need for a stop gauge. Lower headstock unit until with full reach drill will penetrate required depth. Raise up pinion lever, insert work and drill. This is useful when drilling a number of holes to identical depths, as for

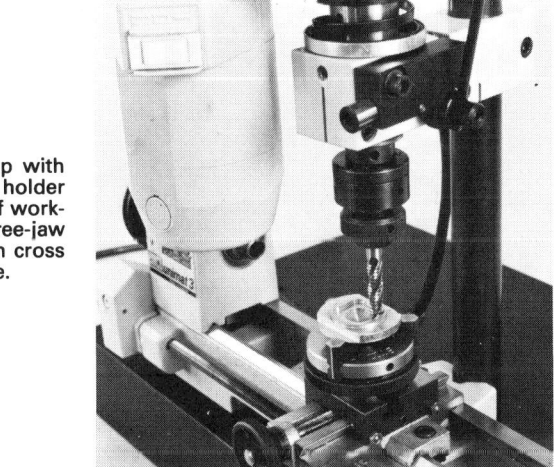

Fig. 10.7 A set-up with reamer in collet holder clearing corners of workpiece held in three-jaw chuck mounted on cross slide carriage.

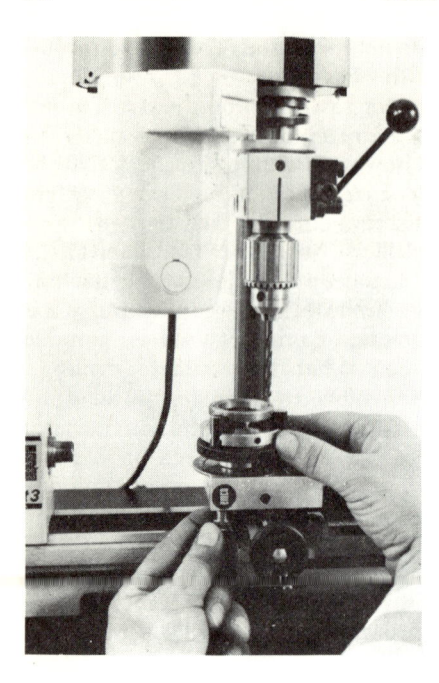

Fig. 10.6 Workpiece in three-jaw chuck mounted on indexing attachment of Mark 3 in cross slide. Evenly spaced fixing holes are being drilled.

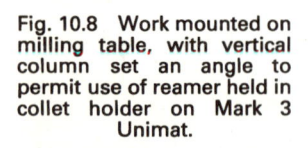

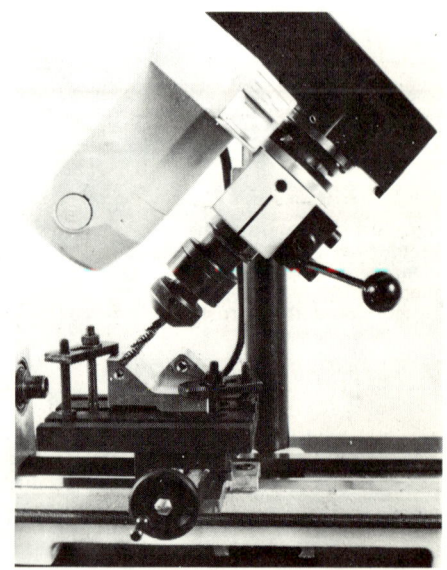

Fig. 10.8 Work mounted on milling table, with vertical column set an angle to permit use of reamer held in collet holder on Mark 3 Unimat.

THE LATHE AS A DRILLING & MILLING MACHINE 91

example in making a drill stand. (Alternative is to drill all holes full depth of material and then add a bottom plate!)

From straightforward drilling one moves naturally to use of the vertical drill for milling purposes. A lot of milling can be done in the lathe, but unless you have made up a vertical slide it is much easier to do it with the work held in the horizontal position. Whether you are end milling or side cutting there is considerable strain on the tool and thence to the headstock spindle to push or bend the equipment out of line. Therefore the shortest protrusion practicable of the tool should be the constant aim, thus reducing the fulcrum action to a minimum.

Quite considerable lengths of slot can be cut by making full use of the longitudinal movement of the slide; cross ways movement is much more limited.

There is a wide range of cutting tools that can be used. We have a fondness for the dental burrs that are so formidable when in the dentist's chair. He, kind man, does not use them to destruction. Make friends with him and beg his part—used throw aways. I have a fine collection from that source! Broken twist drills can also be ground up to make excellent milling cutters by putting an edge to their two flutes.

Fig. 10.9 Ambitious work! Flywheel of Stuart No. 10 vertical engine being drilled to take fixing screw on boss. The Stuart No. 10 can be machined *entirely* on the Unimat Mk. 3.

XI. SIMPLE ACCESSORIES TO MAKE

One of the great problems during all the years that the Unimat has been available has been the absence of regular contributors to the model magazines on its use, on the things that it can be used for to make, and on the design and fabrication of useful accessories that have not been on offer by the manufacturers or might usefully and more cheaply be made in the workshop. The joy of enthusiasts can well be gauged then when quite out of the blue came a certain Mr. Tingey who had exactly the right attitude to the little fellow, plus a considerable inventiveness and the skill to put over his numerous practical accessory ideas. Also I am happy to add his ideas have been very much geared to the capacity of the lathe, employing such materials as light alloy when more orthodox people would have boggled at the thought and urged some tougher, more expensive, less easily worked material. So hats off to Mr. T., who has now produced so many accessories that I can only say thank you, and offer my own small efforts with due acknowledgment to the source!

Somewhere along the line the manufacturers have I fear slipped up a little in their publicity programme. When I visited the factory I came quite by chance on details of a tape recorded series of lessons for handicraft teachers as part of a basic course for the Unimat user. These were in several languages and have proved very successful—indeed a colleague came back from a holiday in Canada with great excitement over a session he had attended with a whole chain of what he called "little lathes" in a class with the teacher overseeing a kind of engineering version of the now well known language lab. On investigation it proved to be the Unimat course I had discovered in Austria. To the best of my knowledge it has never been offered in the U.K., or if so, unobtrusively enough to have been overlooked.

Diagrams in set-ups in their basic instruction manual are so very good that I have re-constructed some for illustration in this book——they can hardly be bettered. Likewise I have made full use of their line drawings with their kind permission.

SIMPLE ACCESSORIES TO MAKE 93

Fig. 11.1   A demonstration class showing students under instruction with the aid of tapes and programmed work. The instructor is just assisting a pupil to set up his vertical column for drilling.

Fig. 11.2   Another stage in audio-visual instruction. Young pupils wear headphones to follow basic instruction course. Note the handy work trays alongside with cutouts to take tools and accessories—well worth copying for the domestic workshop.

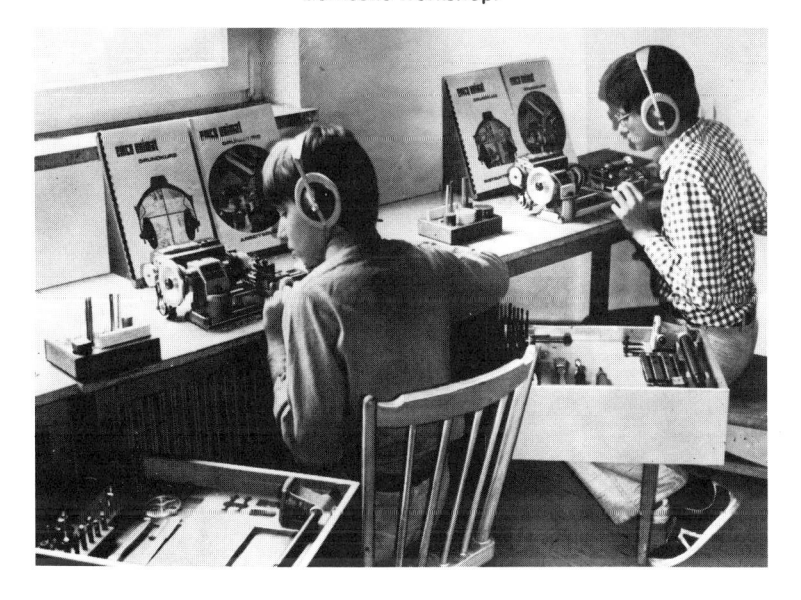

Very much the simplest thing the beginner can attempt is a scriber. This can be elaborated to give practice in facing, drilling, taper turning, recessing and grinding. They are cheap to buy and nothing much will be saved in making your own, but it is the sort of thing that junior members could be turning out rapidly at a club or school open day to impress visitors and raise some funds. At one such function boys were making pokers in this fashion but it seemed to be using an awful lot of metal and their lathe was somewhat bigger than the Unimat. Final thought on scribers—make them of hex bar rather than round if possible—they do not roll so easily!

Vertical Slide

The manufacturers have not as yet put on the market a vertical slide accessory which is a great pity since it can be so very useful enabling work to be done with the tool set up as a lathe instead of as a drill which never seems quite so workmanlike.

I am lucky to have an older second Unimat which is used mainly

Fig. 11.3   Where even slower speeds required than on slowest power setting needs must fit a handle and work the lathe like a sewing machine! This extra has proved very valuable in the hands of *Model Engineer* Editor who devised it for work on iron castings.

SIMPLE ACCESSORIES TO MAKE 95

Fig. 11.4 Vertical slide using discarded cross slide from an earlier lathe. It is attached to cross slide carriage with Dexion angle here, later displaced by stouter angle iron.

as a saw table or grinder where the normal cross-slide is not required. The cross-slide was therefore removed—quite a job in itself since the slide bars must be unscrewed from the bed and the whole lathe casting must be removed from its base since in my older model the bars are secured from underneath. Anyway with the cross-slide released it must be taken to pieces still further by sliding out of its slide bars—they may need a little drifting as they are an excellent fit—and unscrewing the threaded spindle. This leaves the casting free of any parts. Remove and put safely the small securing socket head cap screw.

Next operation in this cannibalising is to cut off flush the socket for the tightening screw. The casting is in a mazak type of material and this represents no problem. Since we are going to stand it up vertically next to the existing cross-slide it must be free to travel along the bed. A short length of $1\frac{1}{2} \times 1\frac{1}{2} \times \frac{1}{8}$th. inch angle iron is cleaned up and checked for true 90° and drilled to take a suitable screw and T-nut to secure to cross-slide. Two holes are drilled in the cannibalised part to secure this to the angle plate. Holes are countersunk to be flush and so not impede action of the slide and suitable bolts put through to be fastened with wingnuts to the angle iron. Put these bolts in place now and reassemble the carriage (they are hidden by this so must be put in first). Put on the wing nuts. Cut down the socket head cap screw already put aside to form part of the T-bolt holding angle iron to the original cross-slide. Take a square nut of appropriate thread and put in 3-jaw. Using a standard T-nut as model, turn down the nut so that it is nicely shouldered and just the right depth to go in the T-slot.

Vertical slide is now complete. Machine vice can be fitted theron, or the milling table (taking care that it clears motor). To give more room the tailstock can be removed from the lathe bed.

Tailpost Dieholder

This one is not original. Credit to Mr. Tingey already mentioned. Novelty here is idea of making it in light alloy. As a dyed-in-the-wool 'strong man' I used to believe no light alloy could possibly be used for things of this sort! I am converted and share the view that the hobbyist uses his tools so little compared with the all-day professional user that dural and aluminium can safely be employed. I am encouraged in this by the increasing use of such materials in photographic accessories—I have yet to find one fail because of the softer material.

At one time all firemen's hose couplings were of solid brass and very heavy. Nowadays they have been largely replaced with lighter and nearly equally serviceable dural; certainly many continental brigades have been so equipped for thirty years or more!

Check size of any die blanks you may have for tapping. Size varies according to diameter of thread to be cut. Metric dies are also very slightly smaller than imperial. Allow $\frac{3}{16}$ in. to $\frac{1}{4}$ in. all round for size of dieholder, which means that for the normal $1\frac{3}{8}$ in. diameter die you will need bar of $1\frac{1}{4}$ in. diameter or slightly more if you are nearer an inch in diameter. You will need a length of about $\frac{3}{4}$ in., but take a little more to allow for chucking.

Chuck and drill through progressively to $\frac{1}{4}$ in. diameter hole. Then drill further with $\frac{5}{16}$ in. drill your drill may have to be stepped to go in $\frac{1}{4}$ in. chuck, but that end can usually be turned down a little as it is not so hard as the drilling end. If you have the Mk. 3 with larger drill chuck you will not have to worry! Then try your hand with the boring tool. It will just go in at this size, and should be carefully positioned so that its rear face—non cutting—does not scrape against opposite side of the hole. It will probably be necessary to pack up the end nearer the work with a shim of about $\frac{1}{64}$ in. thick, or use a piece of visiting card. Adjust so that point and not the whole face of tool is cutting. Speed 2 about 685 rpm should serve, or have it a little faster if you prefer. Bore right through to a diameter of 11

SIMPLE ACCESSORIES TO MAKE

mm. Check this with inside calipers or your Columbus gauge.

Then comes the exciting part, cutting a thread with the threading device. Use the 1 mm pitch leader and follower (quite the most useful size to have) and the inside boring tool. If you have not previously used this equipment much have a trial run first to get the feel of it. It feeds in much faster than you might think! It will need several runs to get a good deep thread cut. Test it with the securing bolt from the grinding wheel set. When it is cut, and before screwing on to the headstock spindle, drill a small hole in the outside diameter just to take the standard tommy bar about $\frac{3}{16}$ in. deep. This will enable you to unscrew your workpiece from the spindle without having to grip it with the wrench!

Now it is safe to slip it on the headstock spindle and give it a final finishing touch with the planing tool. Give the edges of a tiny touch too to chamfer them and take away the sharp corner feeling. All that remains is to drill No. 32 holes to take 4BA grub screws to lock dies in place and adjust gap. There may or may not be locating countersunk pips each side of the slit in the die. The job is now finished and ready for use. There *are* refinements but these are purely cosmetic and can await later experience.

Drill/Reamer Stand

I have already referred to the usefulness of the wood from printers' blocks which come in multi-ply, or, if you are very lucky and get old ones, in beech or other hardwood. I have used them to make stands for drills reamers and other round tools that should not really be left to rub against one another in a box. Normal bought drill stands assume the user will have one only of every drill size—an impossible situation.

It is a good scheme to mark out the block (failing this any flat piece of wood about one inch thick) in say, $\frac{5}{8}$th ins. squares, leaving $\frac{1}{2}$ in. all round; drill at all cross-sections, starting on the front row with $\frac{1}{16}$ in holes, then up either $\frac{1}{32}$nd or another $\frac{1}{16}$ to a row of larger and so on. These rows can then hold drills of the drilled size—which should be about $\frac{3}{4}$ in. deep—with three or four of a size together. Any size you wish to stack but lack a hole for can be used to drill out its own receptable. Size of blocks should not be greater than about six inches long by four deep (these just fit my glass fronted tool cabinet!). They do not get unduly in the way on the bench in this size; can be scrapped without regret when too many size alterations make them a nuisance. Alternately, they can be decently polished up and invoke the envy of less diligent friends. Reamers can of course also be stacked this way: I seem to accumulate loads of them!

Drills are getting expensive, so do not neglect the old dodge with the smaller and more easily broken sizes of drilling out a short piece of rod about ¼ in. deep and slipping the still sharpened tip end into it. Drill slightly oversize, put in a little solder paste (Fryolux or similar) and heat with small blow flame. These short drills are tougher than the original and ideal for drilling shallow holes. The other part of the broken drill can of course be ground to a point and duly sharpened to serve likewise, or with a slight change of course made into a mill.

Slitting Saw Holder

If you have a circular saw then you will probably be able to use the special mandrel for your smaller slitting saws of such great value in metalwork. However, these little saws come in a whole variety of hole sizes—never the size for which you are equipped. They cost very little on the surplus market so that time spent on making a suitable adapter is worthwhile.

You will already have the three parts that comprise the attachment for holding grinding wheels, namely a main part to screw on to the headstock mandrel, a stout washer and retaining screw with hex cap socket. You will need another piece of the alloy bar about 1¼ in. dia. used for the dieholder. Drill out as before but this time to provide a clearance fit for the retaining screw. Then check hole diameter of your saw and step back one face so that the saw will slip

Fig. 11.5 A simple little adapter to enable the grinding wheel arbor to be used for small slitting saws of non-standard hole diameter.

tightly on to it, depth of step being the thickness of the saw, or thereabouts. (If you have several thicknesses of saw with the same hole small paper washers can be cut). On the opposite face an inside step will have to be turned to take the inner lip on the mandrel piece. Thickness of the whole disc should be the same as that of the grinding wheel it is to replace, namely about $\frac{1}{4}$ in.

Assemble the work on the grinding wheel mandrel set—but as yet without the saw. It will be slightly proud of the original parts and the outside diameter can be finish turned, again just rounding off the edges to give a comfortable unsharp feel.

This accessory has the advantage that it does not demand possession of either the threading accessory or facilities for thread-cutting elsewhere.

XII. UNIMAT AS A WOODWORKING TOOL

It will be clear to every user that the Unimat has been designed primarily as a metal turning lathe. However, there is nothing to stop its use as a woodworking lathe in its metal turning form and size. All that is necessary is to remove the longitudinal and cross slides and replace with the swivelling handtool rest. This involves unscrewing the shears from the bed and screwing out the leadscrew to get the slide off. Reassembled small work can be efficiently turned with existing parts, or a special woodturning drive centre can be bought.

Such a setup is adequate if small work such as turning a chess set or the supporting columns of an ornamental bracket clock or model parts for a period ship will be the required output. If larger work such as the legs of a table or chair are in mind then replacement shears are available. They fit on the same base as the original parts and stick out over the end. The tailstock is perched here somewhat

Fig. 12.1 Original jigsaw design fitted to early Unimats. A spring loaded return spring tensioned the saw on its downward stroke and assisted its upward return.

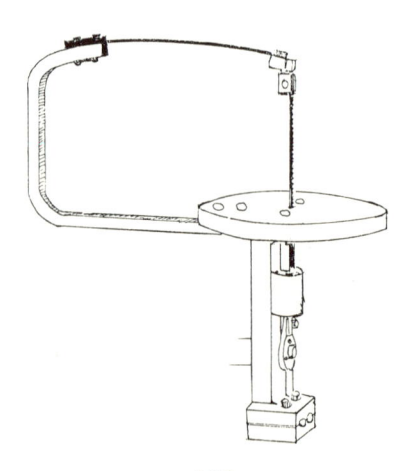

UNIMAT AS A WOODWORKING TOOL 101

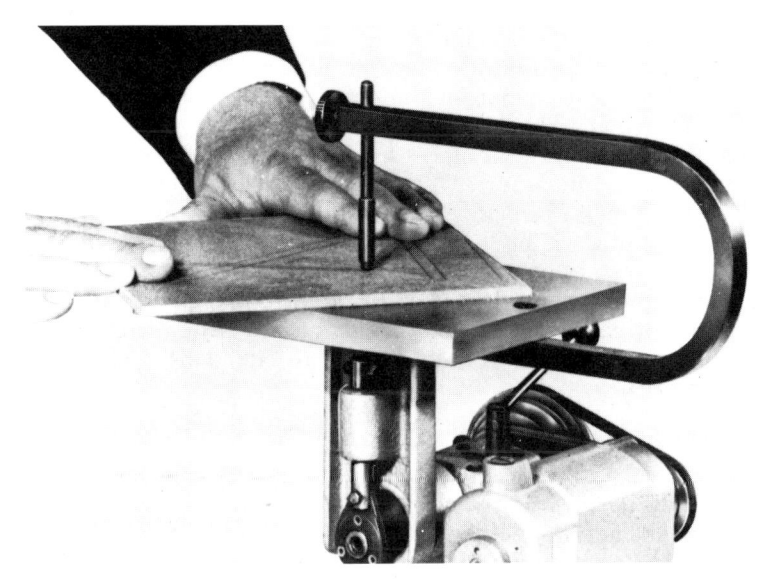

Fig. 12.2 The jigsaw in use. Stout U-shaped support gives reasonable space for work. Saw is supported in channel and help only at the lower end. Channel can be adjusted to allow different thicknesses of wood to be cut.

Fig. 12.3 Sawtable on Mk. 3 set up with sabre saw. Table area is larger than SL size.

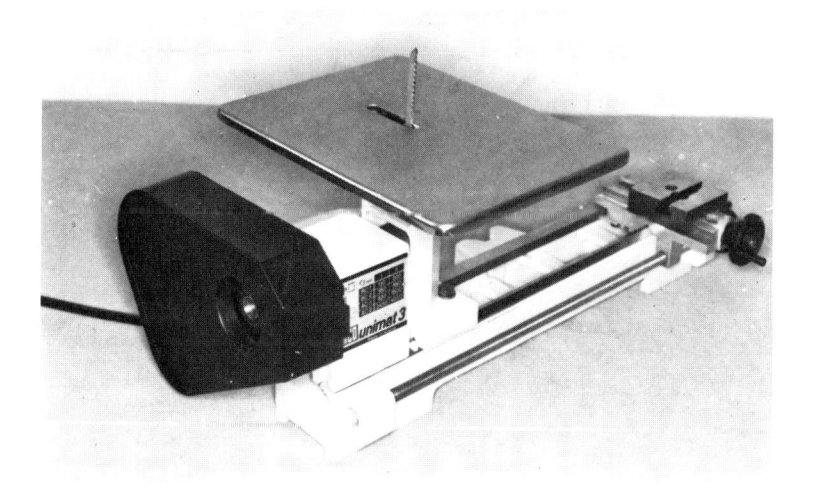

precariously, but the strains of woodturning should not be too great for it; nor will the same high degree of accuracy be demanded. This is the crunch! The equipment is not really suited or intended for this sort of work and like all compromises is less than perfect. Please bear this in mind: if your main use of the Unimat is for small items with only occasional turning, splendid, this will do the work well enough. If you are primarily a woodworker in a big way then do not be unkind to a willing tool—get a separate unit intended for the work you wish to do.

The slightly larger size and considerably more robust nature of the Mk. 3 model make woodworking on it a better proposition, within its centre limits. It is not possible to extend the rigid bed for longer work!

There is a very special pleasure in woodturning. The turning chisels—very like hand chisels though a little smaller—seem to communicate more readily than a metal turning tool. Form tools can easily be filed up from mild steel and will do a lot of work without any need for hardening. The occasional user will find wood a most satisfactory medium for building up patterns that he can later have cast at some friendly local foundry for his metalworking activities. Indeed if it *is* very occasional use there is no need to dismantle the slides—just slip off the tool post and with a little care the slide can support the tool for simple work. For such simple work there is even a gadget that fits onto the longitudinal slide in lieu of the more elaborate swivelling rest. However, it is indeed so simple that a majority of operators will probably make it up for themselves from a piece of scrap material.

The real value of using the lathe for woodworking will doubtless come from the use of such additional equipment as the jigsaw circular saw, sanding plates in both metal and rubber, planing and routing attachments that enable the modelmaker in particular to work on jobs that combine the use of metals, plastics and wood in a general mixture. It is possible to enjoy the fine accuracy of the lathe

Fig. 12.4 Extension piece to raise centre by ¾ in. Note hole has been drilled to go right through so that a longer locating pin can be fitted to ensure firm and accurate setting.

UNIMAT AS A WOODWORKING TOOL

Fig. 12.5 Mitre gauge set to mitre a corner using the circular saw set up on the lathe. It fits in a slotted channel to prevent risk of slip.

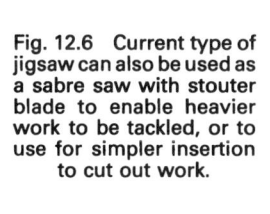

Fig. 12.6 Current type of jigsaw can also be used as a sabre saw with stouter blade to enable heavier work to be tackled, or to use for simpler insertion to cut out work.

Fig. 12.7 Circular saw on older model without raising piece fitted. Large saw table demands that motor be twisted to clear and one pulley must have recess cut in workbench or mounting board to take it. Apart from this and smaller saw diameter it is as efficient as its larger version.

Fig. 12.8 Lathe with extension piece under headstock to enable larger diameter saw to be fitted. Note parallel stop on right of bench.

UNIMAT AS A WOODWORKING TOOL

in producing for example, scale boat gratings from limewood that would be impossible with bulkier 'regular' woodworking machinery.

Certain accessories must be considered with a great deal of planning thought since they make use of only a limited part of the lathe's resources, may require a long time setting up with only limited end benefit, and their need might be better served by another separate instrument. Sawing on the lathe certainly falls into this category.

Here with the jig-saw or sabre-saw only the headstock end and the motor are employed—and the headstock quill serves only to locate the machine, being duly locked fast. This leaves the tailstock and the cross-slide unemployed. In my case I have, perhaps wickedly, relegated my older lathe to act as the woodworking side of the partnership and as such it is from time to time rigged to carry either the jig-saw or the circular saw. In the latter case some use is indeed made of the cross-slide. It serves the humiliating office of locating the accessory equipment.

There are however, some compensations for the non-use of parts if one possesses another lathe. I regularly use toolpost of No. 2 to act, with some modifications as a backtoolpost located on the milling table for parting off purposes. If only one lathe is available I would urge users to give thought to acquiring a separate jig- or

Fig. 12.9 Underneath view of saw table. Sleeve of L-shaped table holder which is attached to cross slide faces downwards in small saw version, upwards with larger diameter and lifting piece.

circular-saw. This may not please the makers but it makes sound sense and maintains the wider usefulness of the equipment.

Earlier versions of the Unimat had an entirely different jig-saw. The table was round instead of rectangular and the upper end of the saw blade was attached to a springy bowlike fixture. The toggle conveying the up and down motion was much less robust than the current model. Today, following the more usual form for light jig-saws the blade is attached only at the bottom or driving end. The upper end runs loose but is constrained within limits by a slotted channel fixed above it and adjustable for height. Since the cut is always on the down stroke and the earlier fixed location did nothing more than hold it in place this can be regarded as a good thing. It certainly means that the slotted guide can be brought down within a minimum distance of the table just allowing enough room for the thickness of the work to pass. This makes for a much truer and more vertical line of cut. It also enables broken blades—heaven forbid!—to be used until they are down to about half-size.

The U-shaped piece supporting the blade guide can be unscrewed and removed when a more robust blade can be inserted and used as a sabre saw without support. This allows unobstructed access to the saw table of quite large workpieces. Should they be excessively large then the screwholes for the removed U-piece can be used to help locate a larger temporary saw table, though it will require additional supports at its corners.

The circular saw set is to some extent a multi-purpose tool. That is to say the saw and its arbor can be used quite separately from the

Fig. 12.10 Useful holding-down clamps of generous size enhance usefulness of Mk. 3 saw-table.

UNIMAT AS A WOODWORKING TOOL

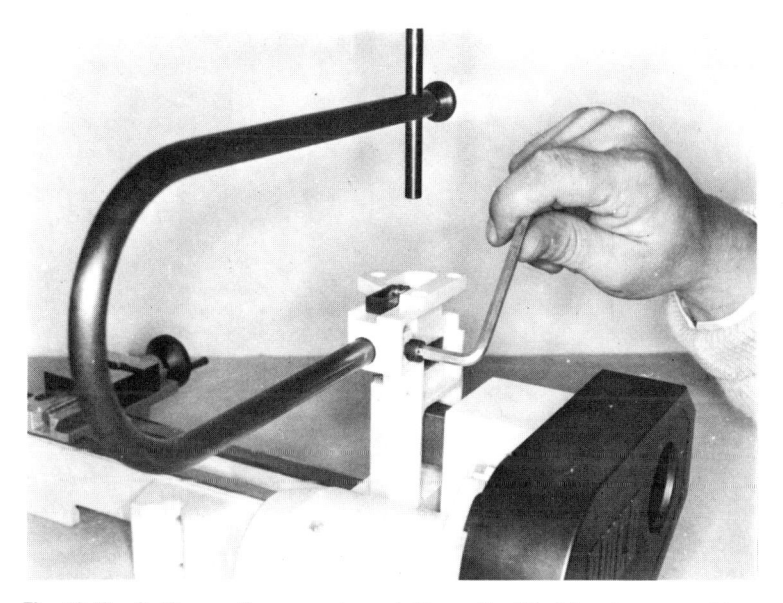

Fig. 12.11 Setting up jigsaw and saw table on the Mk. 3 has been considerably simplified.

saw table for such work as metal slitting when mounted on the headstock set up as part of the vertical drill column. It is a really splendid arbor. A decision must also be taken as to the need for the extension piece allowing a $3\frac{5}{8}$ in. diameter saw blade to be used in place of the normal $2\frac{5}{8}$ in. dia. blade. Since it has so many other uses on the metalwork side it will probably be obtained. Note that the earlier models had a tension screw holding the headstock in place that was tapered and went through a hole in the locating spindle (and in the drill column). This meant that the headstock could not be secured except in its standard position.

Later models have a parallel tension screw which does not go through the column but clicks into one of two circular grooves (also in the drill column) and holds it firmly, either with or without the $\frac{3}{4}$ in. raiser. If you are mean you can of course make this in hardwood or alloy and spend the money elsewhere. Again for the more careful user, remember that small slitting saw blades can often be picked up for pence not pounds in surplus dealers' lists. Against this there may be a wide variety of centre holes requiring special washers to be turned up.

It is important that the blade fits its arbor exactly. For that reason it will probably be necessary to turn down the arbor provided very slightly to fit the official saw blade. Very little has to be turned off,

Fig. 12.12 Wobble-saw usage of saw table for tongue and grooving, with guidebar.

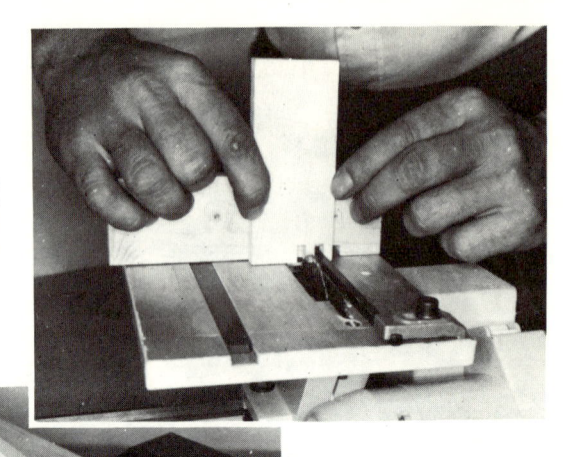

Fig. 12.13 Elegant saw guard on Mk. 3 serves to emphasise thickness of wood being cut.

Fig. 12.14 Using a deeper guard for slotting into end of board.

Fig. 12.15 Mitre gauge in use on Mk. 3 table. This is a compatible accessory.

UNIMAT AS A WOODWORKING TOOL

and blade should click into place. Teeth point in direction of rotation, that is towards you as you face lathe with headstock on your left.

Benefit of the larger diameter blade is largely confined to getting a greater depth of cut. However, it should be noted that the table holder (the bit that clamps onto the cross slide) faces upwards with the larger blade, downwards with the smaller.

Cutting speed should be as high as possible. Recommended speeds:

Wood 1600 rpm	Belt setting 8	
Plastics 850	Belt setting 3 or 9	NB These are conservative speeds & can usually be increased with advantage
Brass/ aluminium 685	Belt setting 2	
Steel 365	Belt setting 1	

A special virtue of the circular saw is its ability to cut an exact straight line (unlike the jigsaw) and so is ideal for picture frames,

Fig. 12.16 (below right) Routing attachment. This is installed at the top of the vertical column with headstock and motor drive underneath. A cutter blade is standard but a round shank type of router can also be used, as may a variety of purpose made form tools.

Fig. 12.17 (below left) Power planing attachment fitted. This is used with headstock in normal position on the lathe bed. Is useful for small precision work such as restoring old wooden cameras and the like.

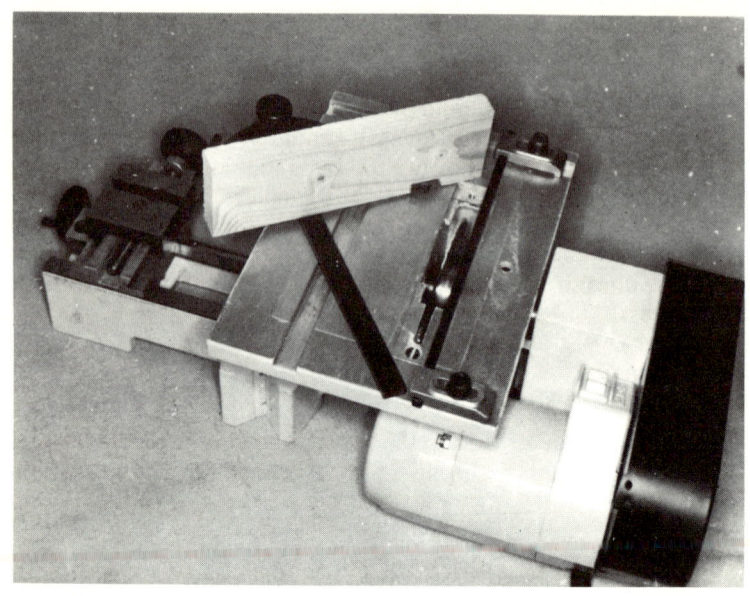

Fig. 12.18 Mitre gauge in use on Mk. 3 table. This is a compatible accessory.

drawers and thin strips for aeromodelling, model boat building. Additional accessory the mitre gauge will be needed to ensure perfect joining for picture frames.

Note that for wood and plastics a fairly coarse toothed blade is required. A finer tooth is used for metals. However, with hardwoods it may be desirable to use a metal type blade to secure the smoothest possible finish.

Two remaining tools of special interest to the woodworker are the planer and router attachments. Since a router usually obtains its fine finish from operating at speeds in the neighbourhood of 20,000 rpm it will be clear that the Unimat's flat out top speed is nowhere near this so that this limitation must be clearly appreciated. I have used a router bit on a power driven hand held drill to rout $\frac{1}{8}$ in. slots in hardboard with quite satisfactory results at a speed not much more than 3,000 rpm so too much should not be made of this reduced speed, provided too much is not asked of the tool.

The router attachment is fitted to the top of the drill column together with the headstock motor. The sideways thrust of the cutter is such that, just for once, it is desirable that the lathe bed be clamped firmly to the bench or located so that it cannot dance about—for a little tool it is very active. A straight cutter is provided

UNIMAT AS A WOODWORKING TOOL

of quite stout proportions with two blades which screws onto the headstock spindle. This will simply give a straight cut—if required to rebate then a strip of waste wood must be fed to it to offer the depth required. However, its real use comes from profiled cutters, giving decorative cuts, mouldings and the like. These are readily obtainable and have a standard 'drill bit' shape with a stem or shank that can be used in conjunction with the drill chuck. As with milling cutters work should be fed against the cutter blade. Failing access to a router bit of desired profile these *can* be made up by the user, but since they must be very sharp to be effective this is really a task for an experienced user. The fence can be unscrewed from the table to give better access to the blade or bit, or a temporary fence made up from scrap wood to provide a curved cut. It needs practice and care to cut freehand and the sabre saw or jigsaw is better for this! Never use the standard cutter without the fence—only round routers!

The planer follows a very similar design to the router, except that it is attached to the lathe with headstock and motor in their horizontal position. Cutter blade is similar to that provided for the router but one inch wide. There is a limited degree of adjustment possible to blade height via the clamping ring which secures the table to the spindle sleeve.

Sharpening of both these cutters follows the usual practice of rubbing on an oilstone, much as one sharpens a hand chisel. Profiled cutters will require touching up with an India slip.

XIII. WOODTURNING ON THE UNIMAT

It will be seen that the smaller woodturning chisel and gouge differ from an ordinary chisel in that they have somewhat longer handles than usual and even longer blades in proportion. This is to give additional control when hand held on the rest. Also the chisel is angled to a sharp point; the gouge is a rounded V-shape. Note also that the chisel cutting edge is bevelled on both sides.

Whilst work can be attached to the faceplate and driven via a lathe dog between centres, it is simpler to use a simple prong chuck which both holds and drives the work. A ballbearing live centre is also desirable since wood can best be turned at fairly high speeds —the most the Unimat will deliver. Lubrication of the live centre is recommended otherwise the wood may heat up and burn.

Fig. 13.1 Mk. 3 set up for woodturning with tools laid out; swivel toolrest is already in place as is spur drive centre.

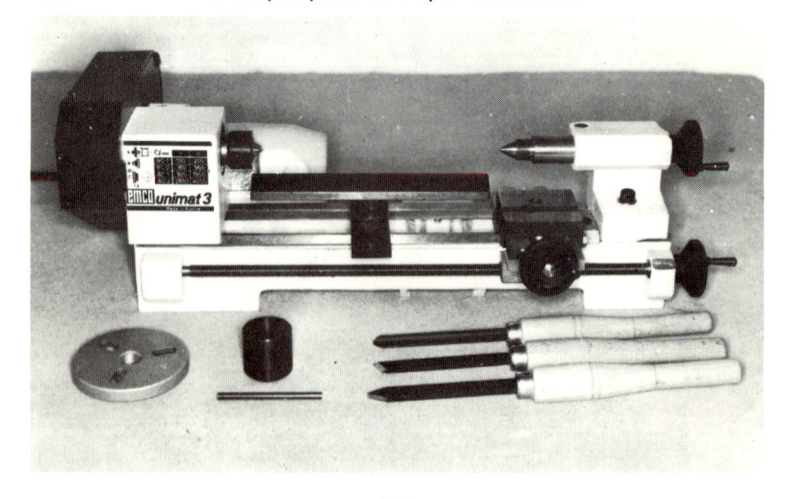

WOODTURNING ON THE UNIMAT

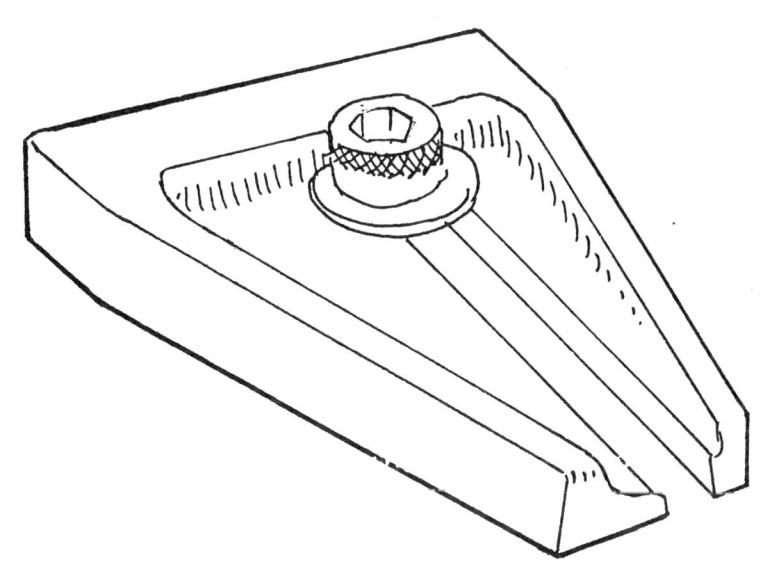

Fig. 13.2 Simple tool rest for the occasional wood turner. This can be bolted on the normal cross slide carriage in place of the tool post. Was originally listed as an accessory but has been dropped in recent lists.

Handrest should be advanced as near the work as possible without fouling revolving parts, adjusted to be at a little below centre height. Tool can be rested on the rest holding it firmly with both hands, at an angle so that the sharp point is clear of the work and to the right, with the rest of the sharpened end facing towards the headstock and working towards it from the tailstock. In working the other way, that is towards the tailstock the blade is reversed

Fig. 13.3 Swivel hand tool rest which is attached to round shear of lathe. It offers a wide range of adjustment for those using the lathe extensively for woodworking.

Fig. 13.4 Making the most of available centre height. Toolrest in position to allow finish turning of small bowl

using its other bevelled edge side. Keep the point out of the way or it will dig in.

The gouge is used initially for the roughing out process when starting a piece of work. This may well be a billet of wood which has just had its four corners planed off to make an octagonal shape, so it will be making intermittent cuts. Do not be too ambitious, gentle small cuts are best; advancing along the work steadily and holding

Fig. 13.5 Standard centre and ball bearing centre on left with spur drive for woodworking which screws directly on to spindle. Next in centre is centring drill and on right faceplate dog with securing screw.

WOODTURNING ON THE UNIMAT 115

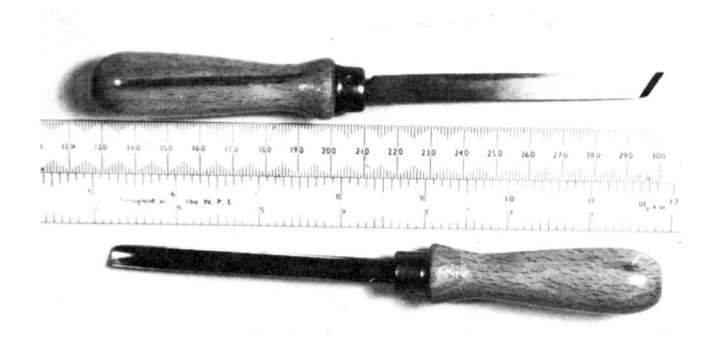

Fig. 13.6 The two small woodworking chisels, skew pointed and gouge can be seen as not quite seven inches long. They are indeed miniatures of the larger standard size tools used on larger woodworking lathes.

the same cut if possible. Perhaps, as a first trial wood that is already rounded may be a better idea. A piece of broomstick, or thick dowelling is a good test piece since it is already compressed and fairly hard, round and will take the tool happily.

The chisel however is the tool for smooth finishing cuts—but watch out for that point digging in. It can be used effectively but not before some degree of skill has been attained. For much the same reason do not use up choice pieces of wood until you have had a little practice.

The softer the wood the harder it is to get a good finish and the more important to have really sharp tools. Working light balsa wood for example is disappointing with knife tools—the finish can only be attained with the aid of fine glass paper. Glasspaper in its various grades is a most valuable aid to finish with most woods. Ideal wood is a close grained type such as beech; other popular and rewarding woods include walnut, mahogany, boxwood (usually only in small pieces) lime. Interesting and attractive effects can be achieved by gluing woods of contrasting colour together to produce items such as trinket bowls, lamp standard, egg cups, plinths for small trophies and the like.

Quite apart from metal and wood it is almost certain that other materials, in particular plastics will be machined on the Unimat. We are indebted to Messrs. G. H. Bloore Ltd of Stanmore for sight of their very informative sheet on machining rigid thermoplastics. These include most notably the following: Acrylic, Acetal, Nylon, P.V.C., P.T.F.E. and Polycarbonate.

Standard metal working lathe and tools should be used at fast speed and fine feed with plenty of suds to keep work cool or even plain water. Tungsten carbide tipped tools may be used (they do not

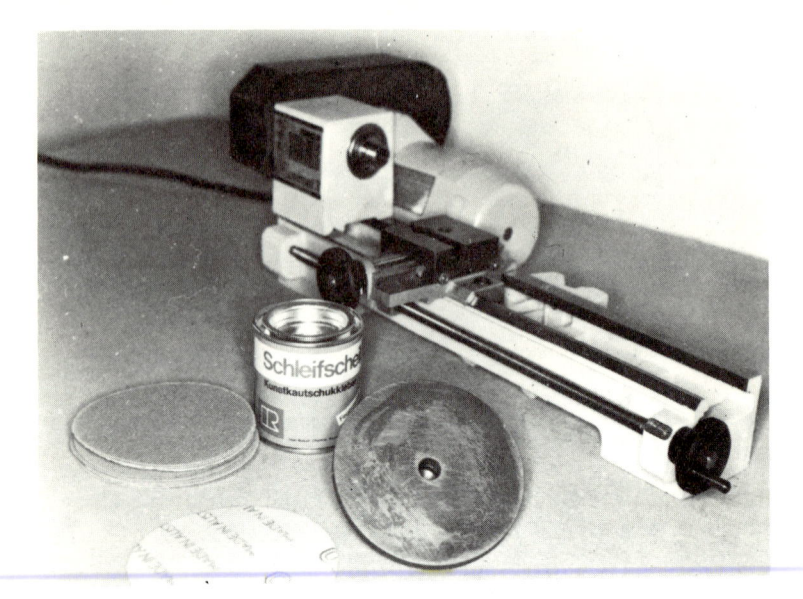

Fig. 13.7 That most useful accessory – the disc-sander being made ready with adhesive to take disc of sandpaper.

lose their keenness when overheated) and sharp tools are highly desirable. The truing diamond used on grinding wheels to restore their surface can also be employed to work plastics, being fixed in the toolpost in the normal way. Recommendations on tool angles suggest a negative top rake from zero to minus 2, front clearance 15°, and side clearance for parting tool 5°.

Since most plastics tend to soften under heat and are in any event less rigid than metal or wood a fixed steady should be used if possible. Lacking this then some form of support pad will help even if made up roughly from wood or other scrap.

It is important that material should not be softened by overheating, therefore working stints should be short and as mentioned above plenty of coolant all the time—though it may be making a mess!

XIV. WORKSHOP, GARAGE, KITCHEN TABLE,...or...

When it comes to a home for the Unimat there are three normal options according to space available. Lucky people, in the minority, will have a separate workshop; less lucky ones will share a limited space with car, motorcycle, mower, even freezer; and the last batch will have to put lathe away each time after use in kitchen, bedroom or living room. Over the years I have had a bit of each of these alternatives.

With a separate workshop, often of wooden construction, a new problem arises, that of preventing rust. If the machine is used almost daily and the workshop is 'lived in' to some extent this is no problem, a simple dust sheet will suffice. With only occasional, perhaps weekend use, then some more positive prevention is required. It goes almost without saying that the equipment is kept reasonably clean and well oiled, with the traditional bottle brush to remove swarf from threads of chucks and mandrel. Something like an office typewriter cover is an excellent idea, using fairly thick material—even cretonne from a loose chair cover to go over the machine—while leaving work in the chuck undisturbed. I do not recommend a plastic covering—it seems to generate a degree of heat, with consequent condensation, whereas the cloth material may get a little damp but absorbs its own moisture. Add to this a bag of silica gel—which is the stuff that usually goes out in the box with a new Japanese camera to perform the same job—and most moisture/rust problems are solved. Your chemist can probably supply as it is also used in laboratory work to maintain a dry constant state—and is very cheap.

Whether in group one or two there is always a need for a workbench. This should be as stout as possible—$1\frac{1}{2}$ in. common deal planks with cross bracing makes an adequate job. Height depends on whether you are a sitter or a stander. With a small lathe jobs tend to take a little longer and I am strictly a sitter. Height therefore should be dependent on comfortable sitting height. A rule over my own work surfaces shows a variation between the three currently in use of 31 ins., 32 ins. and 33 ins. Taking lathe mounting board as 2 ins. thick, this brings lathe centre height up another $3\frac{1}{2}$ ins., so a roundabouts 36 ins. working height of the tool seems to work out.

At the moment I have a corner of garage with bench 37 ins. long and 19 ins. deep which carries the lathe on its board 18 ins. by 9 ins., with hand recesses at each end to lift conveniently like a tray. The bench also shares a robust bench vice on the right. Above in the corner is a corner shelf to take needed tools and support an adjustable lamp which can bend down right on to the work.

This is mainly for dirty work like cast iron, and strictly for winter working. In fine summer weather the trusty Black and Decker Workmate portable bench comes out and has an edged table top which fits snugly on it. An extension power cable comes from the lounge as I work on the patio. In hot weather a nice large umbrella goes overhead. Finally, I am lucky to have an indoor 'office' used variously to write articles, draw, paint, do photography all in a traditional third bedroom space. Here I do 'clean' work on brass, plastics and even a little woodwork. As a friend remarked the only workshop he knew with a carpet on the floor!

Fig. 14.2 Typical small parts nests of drawers that can be easily made up from a few tobacco tins or similar. Drawer knobs can be carpet nails or made up from scrap and dowelling. The keen wood turner could indulge himself with something fancy.

WORKSHOP, GARAGE, KITCHEN TABLE, ... or ... 119

Fig. 14.1 Less than six square feet serve the Unimat as a workshop in a corner of the author's garage. Some form of adjustable lamp to illuminate work brightly is an essential anywhere.

Fig. 14.3 A somewhat larger nest of drawers based on a commercial design. Drawer knobs have been purchased. A metal stationery nest of drawers can sometimes be acquired cheaply or even free if office improvements taking place!

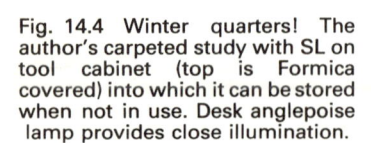

Fig. 14.4 Winter quarters! The author's carpeted study with SL on tool cabinet (top is Formica covered) into which it can be stored when not in use. Desk anglepoise lamp provides close illumination.

Here I use what started life, with space more restricted, as a tidy for photo gear, being a pack flat piece of furniture sold by a mail order house to take gramophone records and the like. This has a top 27 ins. by 14 ins., with the nasty wood veneer now covered with a matching grained sheet of plastic laminate. Inside will take the lathe, tools and accessories and can be packed away in very few minutes. It has casters on the legs but shows no disposition to wander with the Unimat under power. I may take them off and fit domes of silence since it really does not seem quite right! I should add cupboard capacity is 26 in. by $18\frac{1}{2}$ in. by $13\frac{1}{2}$ in., doors are magnetic close with lift-off hinges so that tool array can be readily to hand. Shelves can be arranged to suit needs. This is the ideal for the genuine kitchen table/bedroom model engineer.

WORKSHOP, GARAGE, KITCHEN TABLE, ... or ... 121

Headquarters is undoubtedly the garage and this is the main tool store. On the wall—alas not the one adjacent to the bench—are two shallow glass fronted cabinets each 30 in. by 30 ins., with a depth of just over 4 ins. Ideal not to get in the way of the Mini, but they would have been better six or even eight inches deep. They were made to house small items at an exhibition and can be duplicated with a minimum of trouble since we evolved a design that a holiday-working unskilled student could put together quickly and left him with three dozen to do. With sliding glass fronts they keep all the miscellaneous tools and measuring gear in one place, not swamped by heavier implements and free from dust.

For nuts screws washer bolts and similar treasures I have three other little cabinets. One a six-drawer job, now stained and varnished has been in happy use for ten years or more; the other mere shelvings in scrap wood to hold 4 oz. tobacco tins and a dozen ex-govt tins I picked up ages ago. They were really made to try out a grooving tool I had acquired, but have found things to live in them since. Note the carpet tack handles!

Where to get material and what to do with it is a regular question. The model engineering magazines usually carry advertisements for metal and/or wood in small quantities. Get their lists and order just

Fig. 14.5 Taking up a mere four inches of depth these glass fronted 2½ ft. square wall cabinets (ex-exhibition display cases) house accessories conveniently without getting in the way of the car.

a little more than you need every time. In this way in a few months you will work up a useful stock which will have come to you without financial pain. Keep it on an accessible rack near the work bench.

One of the most useful working aids came quite by accident. I have always had a crude foot on/off switch connected to my photoflood photographic lamps to increase their short and expensive lives. A throw away line in a recent article in *Model Engineer* dealing with Unimat matters had a brief comment something like 'and now is as good a time as ever to make a footswitch' and went on to describe it briefly. Why did I not think of it first? My foot-switch illustrated here is truly a half hour's job and allows use of both hands on the lathe with instant on/off. Since one is so often trying to get more out of the motor than ever intended these off periods will help keep the electrics in high fettle.

Another valuable knock-up that will help protect both lathe shears and chucks is a chuck board. This is a simple piece of wood on two side supporting pieces that just goes across the shears and under the chuck to catch it when being unscrewed from the mandrel. Thus any chance of gashing shears or distorting chuck by a hard bash on metal is avoided. A secondary use for the chuck board is when, very wickedly, one is parting off with a hacksaw it serves to catch the blade when it has gone through the workpiece and would otherwise have added a nasty scratch to the shears!

Fig. 14.6 A foot switch-knocked up in half an hour from a couple of pieces of chipboard, an old hinge and an on/off button from the electrics shop.

WORKSHOP, GARAGE, KITCHEN TABLE, ... or ... 123

Whilst on wooden bits and pieces, if you have a friendly printer acquaintance try and get some of the old wooden blocks on which metal printing plates were fixed (nowadays it is mainly re-usable metal blocks, alas). These are of a standard height, just over half an inch, once of beautiful beech, now mainly multi-ply. They come in very handy suitably drilled out to hold lathe tools, tommy bars and so on during work; or can be drilled out to take taps, reamers, twist drills instead of letting them blunt themselves higgeldy-piggeldy in a tin. At a pinch a wood block can also be brought into service as a fixed steady with the aid of a bit of angle iron.

APPENDICES

TECHNICAL SPECIFICATION

	SL		Mark 3	
	in.	mm	in.	mm
LATHE				
Centre height	$1\frac{5}{8}$	36	$1\frac{13}{16}$	46
Distance between centres	$6\frac{3}{4}$	170	$7\frac{7}{8}$	200
Swing over bed	$2\frac{3}{4}$	70	$3\frac{5}{8}$	92
Cross slide movement	2	51	$2\frac{1}{16}$	52
VERTICAL DRILLING/ MILLING MACHINE				
Max. Working height	$4\frac{3}{4}$	120	$5\frac{1}{2}$	140
Throat	$3\frac{1}{4}$	85		
Drilling stroke	$\frac{3}{4}$	20	$1\frac{3}{32}$	28
Main spindle bore			$\frac{13}{32}$	10.2
Drill chuck . . . up to			$\frac{5}{16}$	7.9
OVERALL DIMENSIONS				
Length	$16\frac{3}{16}$	412	$21\frac{9}{16}$	550
Width	$6\frac{1}{2}$	165	$10\frac{7}{16}$	265
MOTOR				
Capacity		90 watts		95 watts
HP		$\frac{1}{8}$		
Speed range		310/6000*		130/4000
Speed range		365-685-850-2600		130-200-350
(NB Mark 3 has two-speed motor)		3750-6000-1600 1100-200 155*-300*		560-920-1500 2450-4000
THREAD SIZES				
Headstock, tailstock		M12×1		M14×1
Leadscrew		M8×1†		
Socket head cap screws		M6×1		
Hand wheel screws		M5×1		

*plus 155 & 300 with additional idler gear (slow speed attachment)
†left hand thread.

APPENDICES

Summary of principal differences between early/current SL Unimat

Motor
 Wattage increase, progressively over the years from 40 watts—60-75-90 watts.
 On/off switch in lead moved to position on motor.

Headstock
 Division markings added to allow accurate taper turning. Woodruffe key for straight location replaced by locating peg.
 Taper screw fixing (which went right through fixing lug) replaced by straight screw locating in one of two grooves (This enables $\frac{3}{4}$ in. distance piece to be fitted and secured)*

Vertical Drill
 Mazek headstock support re-designed in lighter metal with larger supporting surfaces. Straight screw fixing (see headstock).

Cross-slide base
 Slight casting change to aid swarf clearance.

Lathe Bed
 Shears fixed from top of bed downwards, instead of from underneath upwards, and socket head cap screws replaced by slotted screw heads.

Threading Attachment
 Handle adjustment of threading tool replaced by set screw.

Jigsaw
 Round sawing table replaced by larger rectangular table. Spring return clamp for blade replaced by slotted channel with retaining clamp only at base. This also allows use of sabre saw and is in line with latest modern practice.

Circular Saw
 Introduction of $\frac{3}{4}$ in. raiser block to allow larger diameter saw (and deeper cut)

Ballbearing Live Centre
 Single ballbearing version discontinued
 Double b.b. remains available.

General
 Original bright alloy handles replaced by new design in black. (Cosmetic change only! This style also used on Mk. 3 where it is part of the black/white scheme).

*Note: Height adjuster can *only* be fitted with models having straight fixing screw with two locating grooves, which limits earlier models to the smaller diameter saw.

TOOLS	SL	MK3	ACCESSORIES	SL	MK 3
Assortment of tool bits, ground	—	✓	3-jaw s.c. chuck	✓	✓
2 Thread cutting tools (internal & external)	✓	✓	4-jaw chuck	✓	✓
5 tool bits	—	✓	3-jaw drill chuck	(¼) ✓	(5/16) ✓
Assorted tools (large variety)	—	✓	Vertical drilling & milling attachment	—	✓
ditto (small variety)	—	✓	ditto complete with drive unit	—	✓
Gear Milling Cutter (40 mm dia. module 0.5, No. 1)	✓	✓	Drive unit	—	✓
ditto No. 2	✓	✓	Vertical fine feed attachment	✓	✓
ditto No. 3	✓	✓	Power feed gear (self-act)	—	✓
ditto No. 4	✓	✓	Topslide taper turning	✓	✓
ditto No. 5	✓	✓	Collet holder	✓	✓
ditto No. 6	✓	✓	Set of collets (metric) 16 (0.5-8 mm)	✓	✓
Set of Gear Milling Cutters (dia. 40 mm, module 0.5 No. 1-6)	✓	✓	ditto 10 (Inch $\frac{1}{32}$ in-$\frac{5}{16}$ in.)	✓	✓
Circular Saw Blade for metal (60 mm dia.)	✓	✓	Steady rest	✓	✓
2 circular brushes (brass & steel wire)	✓	✓	Polishing arbor	✓	✓
			Thread cutting attachment	✓	✓
3 felt discs (dia. 40×8, 50 ×10, 70×15 mm)	✓	✓	set of 9 formers & guides (metric pitches 0.5-1.5 mm)	✓	✓

Item		
12 sanding discs (for rubber disc) grit 120	(100) ✓	✓
ditto grit 80	✓	✓
ditto grit 50	(60) ✓	✓
Polishing Lambskin	✓	✓
ACCESSORIES Jigsaw complete with 12 blades	SL ✓	MK 3 ✓
Circular Saw attachment	✓	✓
Adapter to increase centre height	✓	✓
Grinding Wheel guard	✓	✓
Swivel Toolrest (woodturning)	✓	✓
Spur Drive centre (woodturning)	✓	✓
Planing attachment	✓	—
Routing Attachment with cutter	✓	—
Mitre gauge	✓	✓
Watchmaker's collet sleeve	✓	—
Arbor for module cutter	✓	—
Return spring with spring cap for quill (drilling)	✓	—
Works clamps (pair)	✓	
Live centre	✓	✓
Clamping plate	✓	✓
Milling table	✓	✓
Machine Vice	✓	✓
Indexing & dividing attachment (with 24 division plate)	(48) ✓	✓
Division plate 30 div.	✓	✓
ditto 36 div.	✓	✓
ditto 40 div.	✓	✓
Cutter arbor	✓	✓
Flexible shaft	✓	✓
Rubber disc for flexible shaft	✓	✓
Mounting bridge (for milling between centres)	—	✓
Tool grinding attachment (inc. grinding wheel)		✓
Fly Cutter (with 1 tool ground, 2 unground, operating key)	—	✓
Slide unit for milling	—	✓
Safety glasses	✓	✓

BIBLIOGRAPHY

Operating Instructions for the Unimat Universal Machine Tool (3rd Edition)
Operating Instructions for the Emco-Unimat Model SL Small Machine Tool
Emco Unimat Basic Course Instruction Manual
Emco Programmed Exercises Emco-Unimat Small Universal Machine Tool
Edelstaal Unimat Miniature Machining Techniques
The M.E. Lathe Manual *by Edgar Westbury*
Metal Turning Lathes *by Edgar Westbury*
Milling in the Lathe *by Edgar Westbury*
The Amateurs Lathe *by L. H. Sparey*
The Beginners Guide to the Lathe *by Percival Marshall*
Sharpening Small Tools *by Ian Bradley*
Lathe Accessories *by Edgar Westbury*
The Amateur's Workshop *by Ian Bradley*
The Beginner's Workshop *by Ian Bradley*
Myford ML7 Manual *by Ian Bradley*
Myford ML 10 Manual *by Ian Bradley*
Using the Small Lathe *by L. C. Mason*
Making the Most of the Unimat *by Rex Tingey*
Know Your Lathe. . Boxford/MAP
Practical Mechanics Handbook *by F. J. Camm*
Model Engineering Practice *by F. J. Camm*
In the Workshop Vols I, II, III *by Duplex*
Workshop Equipment *by Bradley & Hallows*

All published work by LBSC, Martin Evans, Henry Greenly

All back numbers (and current issues!) *Model Engineer*, 1898 to date
Any back numbers procurable of *English Mechanics, Mechanics, Popular Mechanics, Practical Mechanics.*

There are lots of other useful books of course, I am quoting mainly from my own library or books to which I have access. I never pass a second hand bookshop, or 'cheapie' tray without a quick look through—there are many reading gems yet to be gathered in. Your local library will probably have many of these and some new titles. Keep a jotting book and note down any clever way of doing things—it will come in useful sure enough!